Convex Analysis: Theory and Applications

Translations of

MATHEMATICAL MONOGRAPHS

Volume 222

Convex Analysis: Theory and Applications

G. G. Magaril-Il'yaev
V. M. Tikhomirov

Translated by
Dmitry Chibisov

American Mathematical Society
Providence, Rhode Island

Г. Г. Магарил-Ильяев, В. М. Тихомиров

ВЫПУКЛЫЙ АНАЛИЗ: ТЕОРИЯ И ПРИЛОЖЕНИЯ

УРСС, Москва, 2000

Translated from the Russian by Dmitry Chibisov

The present translation was created under license for the American Mathematical Society and is published by permission.

2000 *Mathematics Subject Classification.* Primary 52–01, 46N10; Secondary 47N10.

For additional information and updates on this book, visit
www.ams.org/bookpages/mmono-222

Library of Congress Cataloging-in-Publication Data

Magaril-Il′yaev, G. G. (Georgii G.), 1944–
[Vypuklyi analiz teoriia i prilozheniia. English]
Convex analysis : theory and applications / G. G. Magaril-Il′yaev, V. M. Tikhomirov ; translated by Dmitry Chibisov.
p. cm. — (Translations of mathematical monographs, ISSN 0065-9282 ; v. 222)
Includes bibliographical references and index.
ISBN 0-8218-3525-4 (acid-free paper)
1. Convex geometry. 2. Discrete geometry. 3. Functional analysis. 4. Operator theory. I. Tikhomirov, V. M. (Vladimir Mikhailovich), 1934– II. Title. III. Series.

QA639.5.M3413 2003
516.3′62—dc22 2003062858

10 9 8 7 6 5 4 3 2 1 08 07 06 05 04 03

Contents

Preface

Convex analysis is a branch of mathematics that studies convex sets, convex functions, and convex extremal problems.

Historically the beginnings of convex analysis date back to antiquity, but its name was coined only in the 60s of the 20th century. Many remarkable facts of convex analysis have been obtained quite recently. So, this is at the same time an ancient and modern part of mathematics.

Convex analysis is closely related to geometry (in which it has its origins, since convexity is a geometric notion), but it is also deeply connected with analysis, and these latter connections stimulated intense interest in convex analysis in recent period. Hence the duality in the very name: *convex analysis.*

At first sight it may seem surprising that convex functions and convex sets, which are very special objects of analysis and geometry, have diverse applications in mathematics, mathematical physics, technology, and economics. In fact, there are good grounds for that (which will be exposed in this book), but anyhow, the fact that the theory of convexity has diverse and fruitful applications is now so unquestionable that basic knowledge of convex analysis is nowadays necessary for almost every mathematician (especially for those dealing with applications). Moreover, studying convex analysis can be motivated aesthetically because it comprises many beautiful phenomena and facts.

Apparently, certain elements of the convexity theory should become part of mathematical education at any level. The present book is intended to contribute to this.

Convex analysis is one of the disciplines which require little preliminary knowledge for their study. This book was also written for a broad readership. Specially for "beginners" we have written the Introduction, where the fundamentals of convex analysis are illustrated in an elementary finite-dimensional setting.

Moreover, the subject matter of Chapter 1 dealing with theoretical aspects of the convexity theory also relies only on the notion of n-dimensional space and therefore, we believe, is accessible to a fairly broad readership.

The geometrically minded reader is advised to combine reading with drawing, since many concepts and proofs have a very simple (and beautiful) geometric interpretation (which we tried to demonstrate by our figures).

This book was originally conceived as the translation of the Russian publication by Magaril-Il'yaev and Tikhomirov (2000), but in the course of preparation of the English edition many parts were essentially revised, so that it became actually a different book. For this reason we somewhat changed its title.

We express our deep gratitude to our colleagues, J. Brinkhuis, A. D. Ioffe, C. V. Konyagin, V. L. Levin, K. Yu. Osipenko, and V. Yu. Protasov for the material kindly provided to us, for useful discussions, and for assistance in preparation of the Russian edition of this book. We thank the student S. S. Chudova for drawing the figures. We are grateful to the translator, Professor D. M. Chibisov, for competent advice and comments which led to improvement of the book.

Introduction

1. What is the subject of convex analysis?

Recall that a set in a plane (or, generally, in a vector space) is said to be *convex* if, along with any two its points, it contains the entire line segment joining them; see Figure 1.

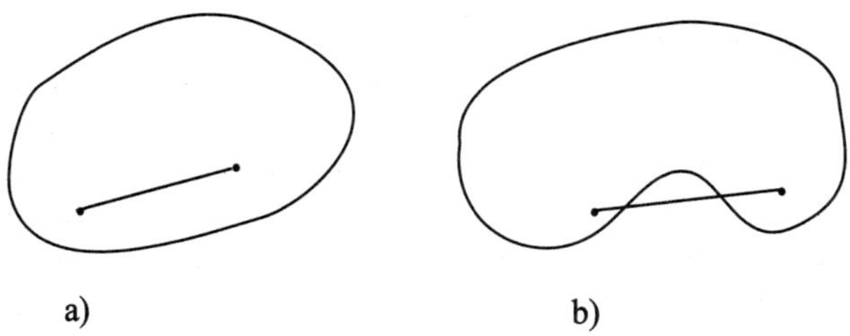

FIGURE 1. a) Convex set. b) Non-convex set.

The boundary of a planar convex set is a *convex curve*, and in the case of a higher dimension it is a *convex surface.* A real-valued function defined on a real line or on a vector space is a *convex function* if its *epigraph*, i.e., the set lying over its graph, is convex (or, in other words, if the line segment joining any two points of the graph lies on or above the graph); see Figure 2.

Seeking the minimum of a convex function on a convex set is referred to as a *convex extremal problem* or a *problem of convex programming.*

Convex analysis is the branch of mathematics which studies convex objects, i.e., convex sets, convex functions, and convex extremal problems.

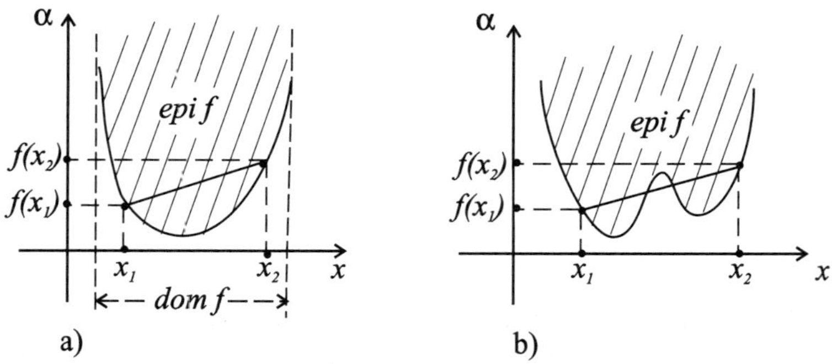

FIGURE 2. a) Convex function. b) Non-convex function.

2. Brief historical review.

The concept of convexity appeared in antiquity. In his treatise "On the Sphere and Cylinder" Archimedes wrote (Axiom 4): "I call convex in one and the same direction the surfaces for which the straight line joining two arbitrary points [...] lies on the same side of the surface." It is the same definition as we gave above (with a slight modification: our definition involves a "line segment" rather than a "straight line"; no doubt, this is what Archimedes meant). Surprisingly, "our" definition was given only in the 20th century. The first modern definition of the convex set is attributed to the German mathematician E. Steinitz, who proposed it in 1913.

As a separate branch of geometry, convex geometry appeared in the 19th century in the works of Cauchy (who proved, in particular, the theorem on rigidity of polyhedra), Steiner, and Minkowski. Convex geometry as a new field of mathematics takes its origin from the publication of the book by Minkowski (1910). This book influenced the formation of a new field in mathematics, viz., functional analysis.

Convex geometry came into vogue in the 30s. The monograph of German mathematicians Bonnesen and Fenchel (1934) summarized the development in this direction for the preceding century.

In the 40s the theory of convexity experienced a striking surge of interest, owing primarily to the so-called problems of *linear programming* which consist of finding an extremum of a linear function on a polyhedron (specified by a finite system of linear equalities and inequalities). Many problems of this kind occur in economics.

The pioneering role in the development of linear programming as a tool of economic planning was played by the Soviet mathematician and economist L. V. Kantorovich, who has written the first treatise on this subject, Kantorovich (1939). For certain political reasons he was not allowed at that time to continue activity in this area. For many years subsequent chapters of the new theory were being written outside the USSR. An intensive development of linear programming and then convex optimization in a broader sense began in the USA by the end of World War II and right after it. It was stimulated by the problems of economics as well as by military needs, in particular, the problems of governing the military forces. Among many researchers who contributed to the rapid development of linear programming and especially its economical applications, we should mention the outstanding mathematicians L. V. Kantorovich and J. von Neumann as well as the great economists W. Leontief and T. Koopmans. (For the elaboration of the mathematical foundations of the theory of economics, Kantorovich and Koopmans were awarded the 1975 Nobel Prize for Economics.) The *simplex method*, the powerful algorithm for solving the linear programming problems which was introduced by the American mathematician G. B. Dantzig, greatly enhanced the practical applicability of linear programming. The book Dantzig (1963) is the first fundamental monograph on linear programming.

In 1949, Fenchel (who had emigrated to Canada) gave a course of lectures which laid the foundation for the theory of convex functions. His monograph Fenchel (1951) stimulated the further development of convexity theory. In two decades Rockafellar (1970) set out recent progress in this direction. The title of this book, *Convex Analysis*, was the first appearance of this combination of words. In the Preface the author thanks Professor A. W. Tucker of Princeton University, saying that it was he who proposed the title of this book. Since then the term "convex analysis" has become generally accepted.

We mention also the monographs by Ekeland and Teman (1976), presenting many applied aspects of convex analysis; and Ioffe and Tikhomirov (1974), containing an introduction into convex analysis and giving a detailed account of its use in extremal problems.

In subsequent years convex analysis made further progress, which will be partly covered in this book.

We begin with a review of the basic ideas and principles of convex analysis; along with this we will set out the structure of the book.

3. Elementary convex geometry.

We will review here the key issues of convex geometry with emphasis on their geometric essence.

All the proofs will be carried through on a Euclidean plane in a purely geometric manner, with the aid only of ruler, compass, and imagination. For example, we will not use (except the parts in small print) functions, nor even numbers (although subsequently, beyond this Introduction, the proofs will be mostly algebraic). The proofs we give here are so simple and natural that looking at the figures, one can see the validity of theorems and the essence of the proofs without logical arguments.

For the sequel we need the notion of closedness. The easiest way to define it is to use that of an open set. A set on the plane is *open* if with any of its points it contains a circle centered at this point. A set is *closed* if its complement is open. Note also that any straight line partitions the plane into three parts: the line itself and two (open) half-planes. If we join the line to one of them, this will be a closed half-plane. Henceforth we call it simply a half-plane.

As in school textbooks, we denote the points of the plane by capital letters, while the sets (sometimes called figures) by script style letters.

We begin with the theorem that is fundamental for all convex analysis, namely, the *theorem on strict separation* (also called in the sequel the second separation theorem).

DEFINITION 1. Let $\mathcal{A}$ be a set and B a point in the plane. A straight line $\mathcal{H}$ *strictly separates the set* $\mathcal{A}$ *from the point* B if $\mathcal{A}$ lies in one of the half-planes bounded by $\mathcal{H}$, while B lies in the other one and does not belong to $\mathcal{H}$ (or, in other words, B lies in the other open half-plane); see Figure 3.

THEOREM (on strict separation). *A convex closed set in the plane may be strictly separated from a point not belonging to it.*

PROOF. Let $\mathcal{A}$ be a convex closed set and B a point outside it. From among the points of $\mathcal{A}$ we can find the point C closest to B (see Figure 3).

Let us imagine that $\mathcal{A}$ is an island amongst still waters of a lake, and we drop a stone "into point B" in the lake. At first the circle

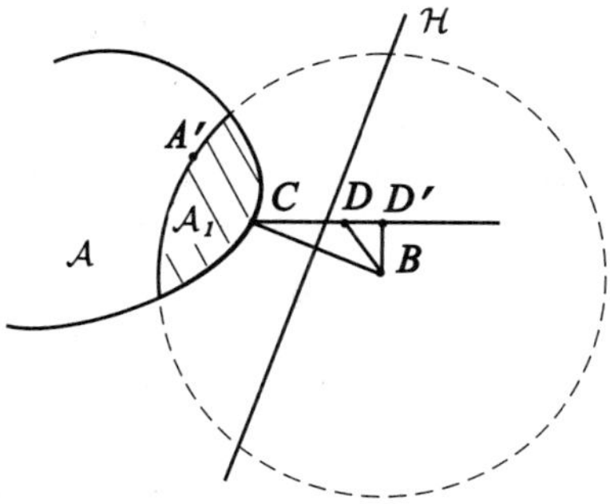

FIGURE 3

extending from this point does not reach $\mathcal{A}$, but at some time the wave front will first touch $\mathcal{A}$ at some point C, which will be the closest point to B.

For a long time mathematicians considered this type of assertion on existence of the closest point as obvious, accepting them without proof. In the 19th century A. Cauchy and K. Weierstrass provided a proof based on the notions of continuity and compactness. We will outline it here (remaining "within the plane").

A real-valued function f defined on the plane is said to be continuous at a point A_0 if for any number $\varepsilon > 0$ there is a number $\delta > 0$ such that the values of f at A_0 and a point A differ in absolute value by less than ε whenever the distance between A and A_0 is less than δ. Weierstrass proved the important theorem that a continuous function on a compact set (in our case on a bounded closed set) attains its maximum and minimum values on this set. Let us apply this theorem to our case.

The distance from a fixed point B to a point of the plane is a continuous function on the plane. Take some point A' in $\mathcal{A}$ and consider the set $\mathcal{A}_1$ which is the intersection of $\mathcal{A}$ and the circle with center at B of radius equal to the distance from B to A' (see Figure 3). By the Weierstrass theorem the distance from B to a point of the plane attains minimum on $\mathcal{A}_1$ (and hence on $\mathcal{A}$) at some point C.

Now the line $\mathcal{H}$ perpendicular to the segment $[C, B]$ and passing through its mid-point solves the problem. Indeed, point B obviously does not belong to $\mathcal{H}$. Assume that the same (open) half-plane where B lies contains some point D from $\mathcal{A}$. Let us drop a perpendicular from B to the segment $[C, D]$ (or the straight line containing it). Denote its base by D'. If D' lies in $[C, D]$, then the length of $[B, D']$ is less than that of $[C, B]$ (as a leg and the hypotenuse of a right triangle). In case D' lies outside $[C, D]$, then BD is less than BC (since the angle BDC is obtuse). Thus we arrive at a contradiction with the fact that C is the closest point to B in $\mathcal{A}$. □

COROLLARY (Minkowski's theorem). *For a figure in the plane to be an intersection of half-planes it is necessary and sufficient that it be convex and closed.*

Indeed, an intersection of half-planes is a convex closed set (because the half-planes themselves are convex and closed). Now let $\mathcal{A}$ be a convex closed figure. Denote by $\mathcal{A}_1$ the intersection of all half-planes containing it. If there were a point B in $\mathcal{A}_1$ but not in $\mathcal{A}$, then separating it from $\mathcal{A}$ would give us a contradiction.

The theorem on strict separation and its corollary were proved by H. Minkowski (1894 – 1909) for a finite-dimensional space and extended to infinite-dimensional case by the Polish mathematician S. Masur (1905 – 1981). This theorem is basic for infinite-dimensional convex analysis.

One of the main theorems of convex analysis is the Minkowski–Krein–Milman theorem on extreme points.

DEFINITION 2. The points of a set that are interior for no interval contained in this set are said to be *extreme points* of this set.

The extreme points of a triangle are its vertices; for a disk these are the points of its boundary circle (Figure 4).

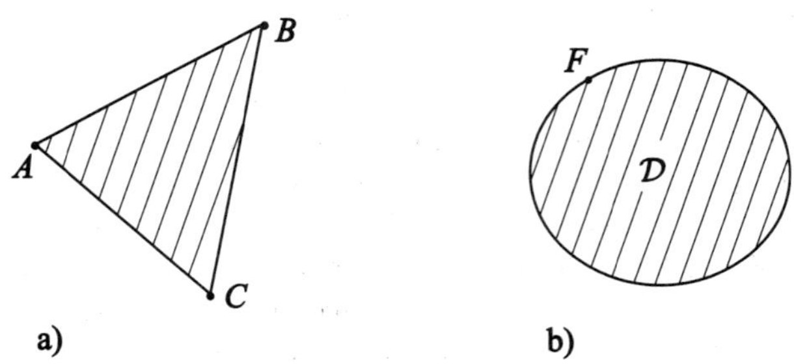

FIGURE 4. a) A, B, C are the extreme points of the triangle ABC. b) F is an extreme point of the circle $\mathcal{D}$.

DEFINITION 3. The intersection of convex (convex and closed) sets containing a given set is its *convex hull* (*convex closure*).

THEOREM (Minkowski–M. Krein–Milman). *A convex bounded planar set is the convex closure of its extreme points.*

PROOF. First of all we prove that the set of extreme points of $\mathcal{A}$ is nonempty.

Let $\mathcal{H}_0$ be an arbitrary straight line. One possibility is that $\mathcal{H}_0$ has some points common with $\mathcal{A}$ and the set $\mathcal{A}$ lies entirely on one side of this line, in which case it is a *support line* to $\mathcal{A}$. Otherwise, if $\mathcal{H}_0$ is not a support line, we will "move" it, keeping it parallel to itself until it becomes a support line (Figure 5a and b).

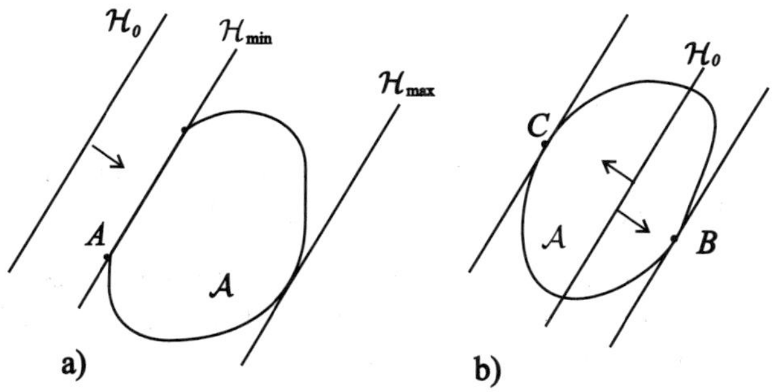

FIGURE 5. a) A is an extreme point of $\mathcal{A}$. b) B is an extreme point of $\mathcal{A}$.

Again, we will explain how this (intuitively obvious) assertion can be proved. Consider the function on the plane, which is the distance from a point on the plane to $\mathcal{H}_0$. This is a continuous function; hence by the Weierstrass theorem it attains its maximum and minimum on $\mathcal{A}$. The parallel lines to $\mathcal{H}_0$ at the maximal and minimal distance from it are support lines.

The intersection of a support line $\mathcal{H}_1$ with $\mathcal{A}$ is an interval (which may degenerate into a point), and the end-points of this interval will be extreme points of $\mathcal{A}$. (Prove it yourself.)

Now let $\mathcal{A}_1$ be the convex closure of the extreme points of $\mathcal{A}$. It is a convex and closed subset of $\mathcal{A}$. Assume that there is a point B in $\mathcal{A}$ but not in $\mathcal{A}_1$. By the theorem on strict separation the point B

can be strictly separated from $\mathcal{A}_1$ by a line $\mathcal{H}_1$. Let us find in $\mathcal{A}$ the most distant point A_1 from $\mathcal{H}_1$ in the same half-plane as B. By the above, the line $\mathcal{H}_1'$ parallel to $\mathcal{H}_1$ and passing through A_1 contains an extreme point of $\mathcal{A}$, which contradicts our construction (Figure 6).□

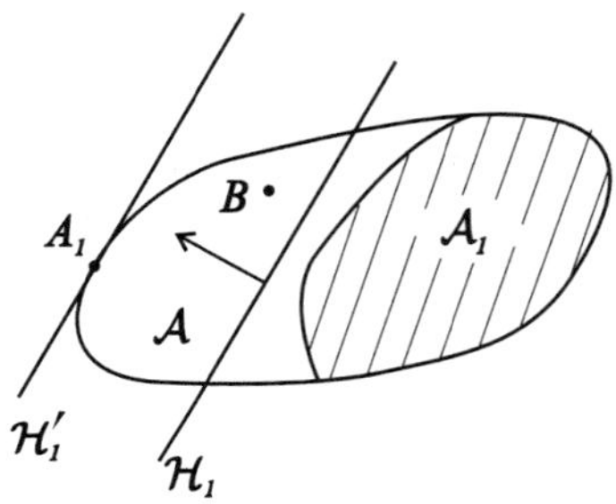

FIGURE 6

EXERCISE 1. Show that the convex hull of a finite number of points is an intersection of finitely many half-planes.

The theorem on extreme points in the infinite-dimensional case was proved by the Soviet mathematicians M. G. Krein (1807–1989) and D. P. Milman (1913–1982) in 1940.

The finite-dimensional setup has a number of features unavailable in infinite-dimensional spaces. One of them (very important in applications) is the possibility of the so-called "clean-up"[1] based on the Helly theorem, which is a remarkable result of convex finite-dimensional geometry. We will prove this theorem here using two results which are of considerable independent interest.

THEOREM (Carathéodory). *If a point belongs to the convex hull of a finite system of points, then it either coincides with a point of this system or belongs to an interval joining two points of the system or to a triangle with vertices in the system.*

[1]This is a literal translation of the term adopted in the Russian literature, meaning that some unnecessary things may be taken away. Specifically, it refers to the possibility of deleting all but a finite number of points, leaving their convex hull unchanged. There is no generally accepted term in the English literature. In the translation of Ioffe and Tikhomirov (1974) the term "decomposition" was used. In this book we adopted the translation "clean-up".

In other words, a point that belongs to the convex hull of a finite system of points belongs to the convex hull of at most three points of this system.

This theorem was proved by the famous German mathematician of Greek origin, C. Carathéodory (1873 – 1950).

PROOF. The convex hull of a finite system of points is a polygon whose vertices are some points of this system (think it over yourself). If the point in question is one of the points of the system or lies on the interval joining two such points, then the proof is over. Assume the contrary. Let us draw a straight line through this point and a vertex of the polygon. It will cross the boundary of the polygon at a point on its side with end-points belonging to the system. Therefore our point belongs to the triangle with vertices mentioned above (Figure 7). □

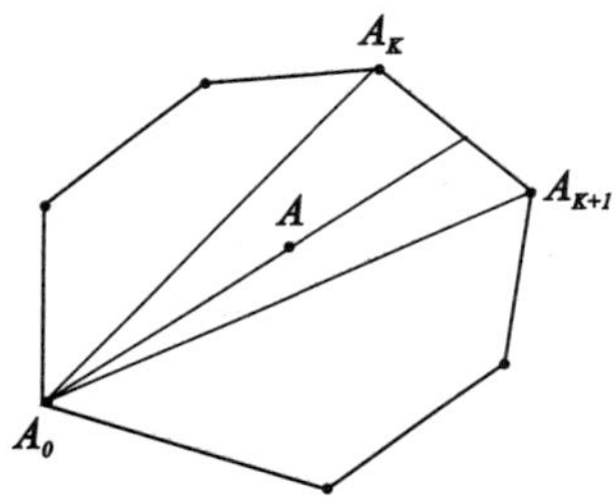

FIGURE 7. A belongs to the convex hull of $\{A_0, A_k, A_{k+1}\}$.

THEOREM (Radon). *Any system of at least four points can be partitioned into two subsystems so that their convex hulls have a nonempty intersection.*

PROOF. It suffices to prove this for a system of four points. Assume that these points do not lie on one straight line. Then their convex hull is either a triangle (see A_1, A_4, A_3 in Figure 8) or a quadrangle (see A_1, A_2, A_4', A_3). In the former case the fourth point belongs to the triangle (the convex hull of its vertices), and in the latter the diagonals (the convex hulls of pairs of points) have an intersection point. □

THEOREM (Helly). *Let a family of at least three planar bounded closed convex sets be given. If any three sets of this family have a*

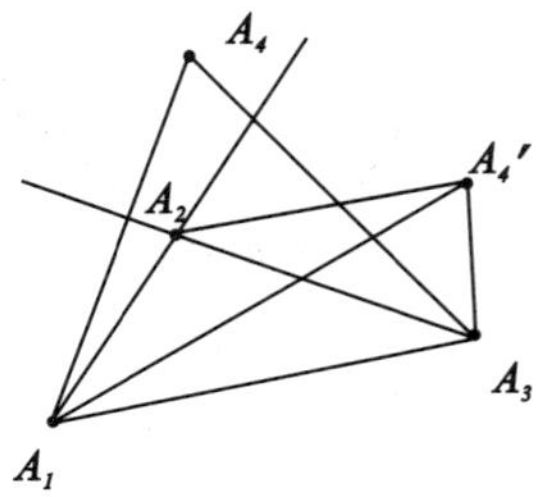

FIGURE 8

nonempty intersection, then there exists a point which belongs to all of the sets of the family.

PROOF. We will prove this statement for a finite system of, say, r sets by induction in r. If $r = 3$, then the assertion follows from the condition. Assume that the theorem holds for $r - 1$ sets, and let a system $\mathbb{A} = \{\mathcal{A}_1, \dots, \mathcal{A}_r\}$ of r sets be given. By condition, for any i $(1 \leq i \leq r)$ there is a point B_i which belongs to the intersection of all the sets of the system except $\mathcal{A}_i$. Since r is no less than four, we can apply Radon's theorem and partition the points $\mathcal{B} = \{B_i\}_{i=1}^r$ into two sets $\mathcal{B}_1$ and $\mathcal{B}_2$ such that their convex hulls have a nonempty intersection. Let B be a point in this intersection. Denote by $\mathbb{A}_1$ and $\mathbb{A}_2$ the corresponding subsystems of $\mathbb{A}$. Each point in $\mathcal{B}_1$ belongs to every set in $\mathbb{A}_2$, hence so does the convex hull of these points. In the same way the convex hull of $\mathcal{B}_2$ belongs to every set in $\mathbb{A}_1$. Therefore their common point B belongs to every set of the system $\mathbb{A}$ (Figure 9). □

EXERCISE 2. Assume that every three of finitely many points on the plane can be placed into a circle of unit radius. Show that all the points can be placed into a circle of unit radius.

EXERCISE 3. Assume that the distance between every two out of a finite number of points on the plane is no greater than one. Show that all these points can be placed into a circle of radius $\frac{1}{\sqrt{3}}$ (Young's theorem on the plane).

The above theorem was proved by the Austrian mathematician E. Helly (1884–1943) in 1913. He spoke about this theorem to his colleague J. Radon (1887–1956), an expert in analysis and measure theory. Helly did not publish the proof of his theorem before World

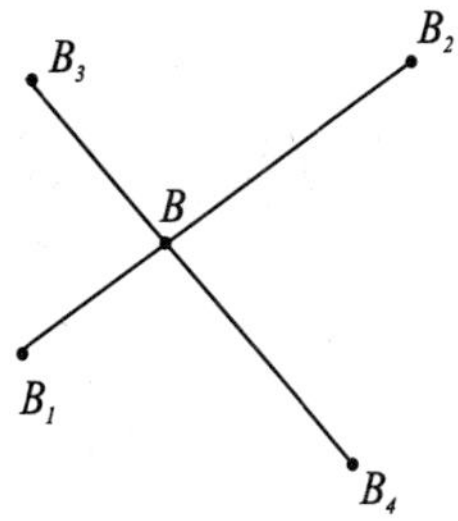

FIGURE 9. B_1 and B_2 belong to $\mathcal{A}_3 \cap \mathcal{A}_4$, B_3 and B_3 belong to $\mathcal{A}_1 \cap \mathcal{A}_2$, hence $B \in \bigcap_{i=1}^{4} \mathcal{A}_i$.

War I. Then he was called up, was wounded at the Russian front, captured, treated in a hospital in Siberia, and returned to Austria only in 1921. His proof of the theorem appeared in Helly (1923). Radon found another proof based on his own theorem, published in Radon (1915).

Now we turn to convex functions. We will consider functions of one (real) variable. Let us draw on our Euclidean plan a horizontal line and a vertical line perpendicular to it. Denote their intersection point by O. One can distinguish between two kinds of straight lines on the plane: the ones intersecting the vertical line and those parallel to it. The nonvertical lines passing through the point O are the graphs of linear functions. Generally a nonvertical line on the plane is a graph of an *affine* function. A set on the plane which with any of its points contains the entire ray going from this point "upwards" parallel to the vertical line is the *epigraph* of the function that associates with any point of the horizontal line the lowest point of the epigraph projecting into this point. The epigraph of a function f is denoted by $\operatorname{epi} f$. A function is said to be convex if its epigraph is a convex set and closed if it is a closed set (Figure 10).

EXERCISE 4. Give an example of a function whose epigraph is convex but not closed.

The following important theorem is one of the fundamental results in convex analysis of functions.

THEOREM (Fenchel–Moreau). *The epigraph of a function is the intersection of epigraphs of affine functions if and only if the function itself is convex and closed.*

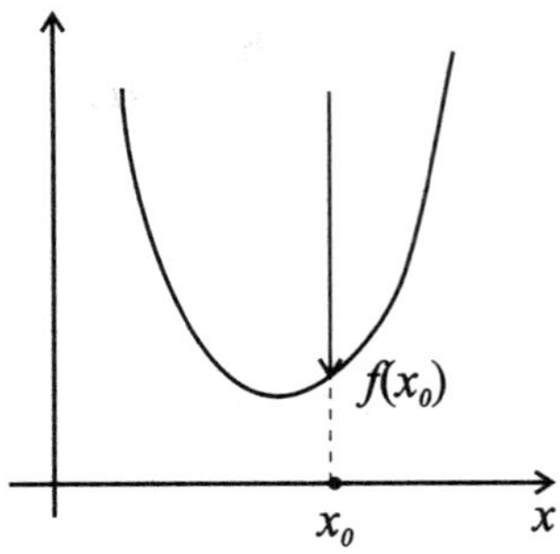

FIGURE 10

This theorem was proved in the finite-dimensional case by the German mathematician W. Fenchel in 1949 (see Fenchel (1951)) and in the infinite-dimensional case by the French mathematician J. J. Moreau in 1960 (see Moreau (1962)).

PROOF. The "only if" part follows from the definitions: if the epigraph of a function is the intersection of epigraphs of affine functions, then the function is convex and closed.

If the epigraph is empty (in terms of functions this means that the function is identically equal to infinity), then the assertion is obvious: take all constants for the affine functions (their graphs are horizontal lines). Suppose the epigraph $\mathcal{A}$ is nonempty and let B be a point on its boundary. Take a point B' "under" B on the same vertical line (Figure 11). The line $\mathcal{H}_0$ strictly separating B' and $\mathcal{A}$ cannot be vertical (because the vertical line through B' passes also through B, contrary to strict separation). Hence it is the graph of an affine function.

Assume now that the set $\mathcal{A}_1$ which is the intersection of epigraphs of affine functions lying below the set $\mathcal{A}$ (clearly $\mathcal{A}_1$ contains $\mathcal{A}$) does not coincide with $\mathcal{A}$. Then there exists a point B_1 in $\mathcal{A}_1$ but not in $\mathcal{A}$. The line $\mathcal{H}_1$ separating B_1 from $\mathcal{A}$ cannot be the graph of an affine function, because that would contradict the construction of the set $\mathcal{A}_1$. Hence it is vertical. The line $\mathcal{H}_0$ cannot pass "over" the point B_1 (which would lead us to the same contradiction). Since the line $\mathcal{H}_0$ is not vertical, it intersects the vertical line. Let C be the intersection point of $\mathcal{H}_0$ and $\mathcal{H}_1$. Take a point B_1' "over" B_1 and draw the line

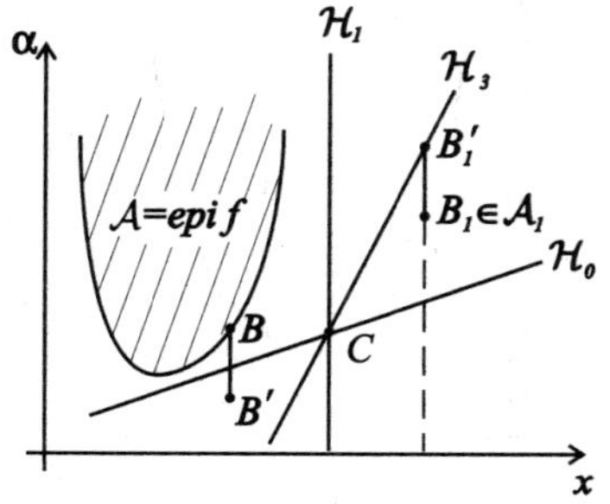

FIGURE 11

$\mathcal{H}_3$ through C and B_1'. It is not vertical and lies over B_1. This again is a contradiction, which proves the theorem. □

4. The main objects of convex analysis.

Besides the two main objects of the convexity theory mentioned above, convex sets and convex functions, we should also mention subspaces, convex cones, convex sets containing the origin (termed *zero-convex sets*), "translated subspaces" (termed *affine subspaces* or *affine manifolds*), and convex homogeneous functions.

Recall that a set in a vector space is a *cone* (with vertex at the origin) if, along with any of its points, it contains the entire ray passing through this point from the origin; a function f is *homogeneous* (of first order) if $f(\alpha x) = \alpha f(x)$ for any positive α. A *convex cone* is a cone which is a convex set, and a *convex homogeneous function* (also called a *sublinear function*) is a convex function which is at the same time a homogeneous function of first order.

5. Duality in convex analysis.

One of the most important phenomena related to convexity is that of duality.

It manifests itself already in terms of the spaces in which our convex objects are considered. Convexity is defined in a vector space, but the theory of convexity takes the most complete form in the spaces endowed with the notion of closedness. This is possible in topological spaces where topology is compatible with convexity, so that the "habitat" of convex analysis where it "naturally lives" is the so-called *locally convex spaces* (LCS) defined in the mid-30s by J. von Neumann (after A. N. Kolmogorov introduced the notion of a linear (nowadays termed "vector") topological space). But in this book we will primarily stay on the "finite-dimensional level" (tackling the infinite-dimensional setting only in Section 12).

The set of linear functionals on n-dimensional space (or linear continuous functionals if the basic space is LCS) forms the dual space.

The fundamental principle of convex analysis may be formulated as follows: *the convex objects (functions, sets, and extremal problems) have a twofold description*; with each convex object in the basic space one can associate the dual object in the dual space; if the initial object has certain closedness properties, it can be uniquely reconstructed from the dual object. We will explain below how the reconstruction is done, and now we give some examples of dual description of convex objects.

To make the discussion more specific, we recall the description of the finite-dimensional space $\mathbb{R}^n$ and its dual (a more detailed description will be given in the next section).

The space $\mathbb{R}^n$ consists of all ordered collections of n real numbers, which we write as column-vectors .

$$x = \begin{pmatrix} x_1 \\ \vdots \\ x_n \end{pmatrix}$$

(but to save space we will also write $x = (x_1, \dots , x_n)^T$, where T means the transpose). For such vectors the operations of (coordinate-wise) addition and (coordinate-wise) multiplication by a real number are defined in a natural way. With these operations $\mathbb{R}^n$ becomes a vector space. As we said, the space dual to a vector space is the space of linear functionals defined on it. If $a = (a_1, \dots , a_n)$, the map $x \to a \cdot x = a_1x_1 + \dots + a_nx_n$, where $x = (x_1, \dots , x_n)^T \in \mathbb{R}^n$ (with the dot denoting the matrix multiplication of the row-vector a by the column-vector x), obviously is a linear functional, and any linear functional on $\mathbb{R}^n$ may be written in this form. Thus the dual space to $\mathbb{R}^n$ may be realized as the set of all ordered collections of n real numbers written as row-vectors. We denote this space by $(\mathbb{R}^n)'$. Let $a \in (\mathbb{R}^n)'$ and $\gamma \in \mathbb{R}$. The set $H = H(a, \gamma) = \{x \in \mathbb{R}^n \mid a \cdot x = \gamma\}$ is a *hyperplane* (so that a hyperplane is a level set of a linear functional). Each hyperplane $H(a, \gamma)$ determines two half-spaces $H_+(a, \gamma) = \{x \in \mathbb{R}^n \mid a \cdot x \leq \gamma\}$ and $H_-(a, \gamma) = \{x \in \mathbb{R}^n \mid a \cdot x \geq \gamma\}$.

Now we can illustrate the basic duality principle. We begin with convex sets.

Any convex closed set in $\mathbb{R}^n$, on the one hand, is a point set which with any two points contains the line segment joining them and, on the other hand, is the intersection of all subspaces containing it (Minkowski's theorem), so that convex sets are solutions to systems of (nonhomogeneous) inequalities. Thus we see that geometry and algebra unite in the description of convex sets.

Similarly, the other objects can also be described in a twofold way. A convex function with closed epigraph (called in this case a closed function) can be dually defined as the supremum of affine functions, i.e., functions which are the sum of a linear function and a constant. The dual description of a sublinear function is the supremum of linear functions. A convex closed cone (with vertex at the origin) is, on the one hand, the union of its constituent rays and, on the other hand, the intersection of the half-spaces with boundary hyperspaces passing through the origin (i.e., cones are solutions to systems of linear inequalities). A zero-convex set is, on the one hand, the union of intervals originating from the origin and, on the other hand, the intersection of half-spaces containing the origin (i.e., such sets are solutions to systems of inequalities with the same right-hand sides).

Affine subspaces are defined geometrically as the sets containing, together with two points, the entire straight line passing through them and, on the other hand, as the solutions of nonhomogeneous systems of equations. Subspaces are affine subspaces which are cones and, on the other hand, are solutions of homogeneous systems of equations.

This is how the unity of geometric and analytic approaches to convexity manifests itself: convex objects are defined both geometrically as sets of points with certain properties and analytically as solutions of systems of linear equations and inequalities.

As we said, the convex closed functions in the dual description are the suprema of affine functions. This leads us to the definition of the dual function in the dual space. In $\mathbb{R}^n$ it is defined as follows. Let $f : \mathbb{R}^n \to \mathbb{R} \cup \{\pm\infty\}$.

The *Legendre–Young–Fenchel transformation* of f or the *conjugate function* to f is the function defined on the dual space $(\mathbb{R}^n)'$ by the following rule:

$$f^*(y) := \sup_x (y \cdot x - f(x))$$

(see Figure 12).

The function conjugate with f^* (defined on $\mathbb{R}^n$) is called *the second conjugate* of f and is denoted by f^{**}.

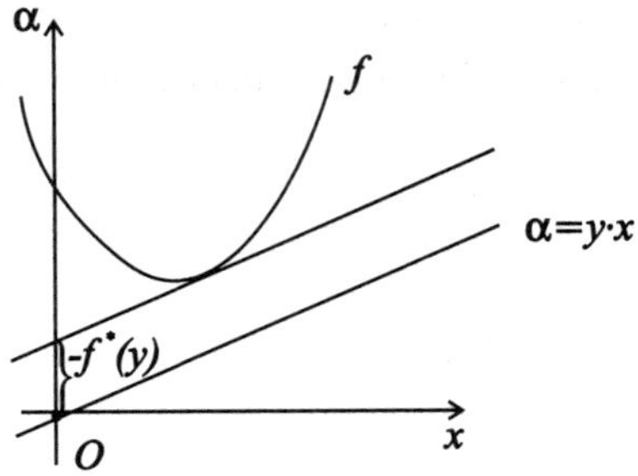

FIGURE 12

The following *Fenchel–Moreau theorem* holds: *if* $f\colon \mathbb{R}^n \to \mathbb{R} \cup \{+\infty\}$, *then* $f^{**} = f$ *if and only if* f *is convex and closed.*

Geometrically this theorem means that a convex closed function is the envelope of a family of affine functions.

This is the fundamental result in the duality theory of convex functions. In turn, this theorem (and actually the entire theory of duality in convex analysis) relies on the *separation theorem*, which was first proved in the finite-dimensional case by Steinitz in 1913, although it had already been known in an implicit form to Minkowski. (It was discussed above for a plane.)

Besides the conjugation operator for functions (which associates with each function its conjugate), there are other duality operators: annihilator for subspaces (which associates with a subspace its annihilator), the conjugation operator for cones (which associates with a cone its conjugate), the polar for zero-convex sets (which associates with a set its polar), and other operators for affine subspaces and convex sets. The separation theorem states that these operators are involutory (i.e., the twofold application of the operator maps the object into itself).

6. Convex calculus.

The calculus of smooth functions (differential calculus) hinges on the calculus of differentials; the main component of convex calculus is the "calculus" of subdifferentials. The subdifferential is an analog of the differential adapted for convex analysis.

To explain this, let f be a function differentiable at a point x_0. Then the differential of f at x_0 is a linear function (i.e., an element of the dual space) which, upon addition of the constant $f(x_0)$, provides a good approximation of f in a neighborhood of the given point.

In convex analysis the role of linear function is played by sublinear functions. Such a function is the supremum of some family of linear functions. The corresponding set in the dual space is referred to as the *subdifferential of the sublinear function*, and each element of the subdifferential is a *subgradient*. Its formal definition is as follows: if p is a sublinear function, then its subdifferential is $\partial p = \{x' \in (\mathbb{R}^n)' \mid x' \cdot x \leq p(x) \ \forall x \in \mathbb{R}^n\}$.

Any convex function f is approximated locally (in a neighborhood of some point x_0) by the sum of a sublinear function and the constant $f(x_0)$. The subdifferential of the sublinear function approximating f at the point x_0 is called the *subdifferential of f at x_0*. It is denoted by $\partial f(x_0)$ and can be defined directly: $\partial f(x_0) = \{y \in (\mathbb{R}^n)' \mid f(x) - f(x_0) \geq y \cdot (x - x_0)\}$ (in this notation the subdifferential of a sublinear function p is $\partial p(0)$). Among the basic theorems of differential calculus there are theorems on the derivative of the sum of functions, on superposition of functions, and on inverse functions. In convex analysis there are analogs of these theorems, as well as results and formulas that have no analogs in differential calculus, e.g., the formula for the subdifferential of the maximum of convex functions.

In particular, under certain assumptions on f_1 and f_2, we have the following formula for the sum (an analog of the corresponding theorem of differential calculus):

$$\partial(f_1 + f_2)(x) = \partial f_1(x) + \partial f_2(x)$$

(the Moreau–Rockafellar formula), and the subdifferential of the maximum of two convex functions continuous at a given point and taking equal values at this point is the convex hull of the union of their subdifferentials; i.e., for the function $f(x) = \max(f_1(x), f_2(x))$ we have the following formula (by Dubovitskii–Milyutin):

$$\partial f(x) = \{y = (1-\alpha)y_1 + \alpha y_2 \mid y_i \in \partial f_i(x), \quad i = 1, 2, \ 0 \leq \alpha \leq 1\}.$$

This result has no counterpart in the classical calculus. At the same time, since convex functions are representable as suprema of affine ones, the formulas for the maxima or suprema of a family of functions play a particularly important role.

Let p be a sublinear function. It is not hard to show that the set ∂p (in the dual space) is convex and closed. Then a natural question is how to reconstruct the function from its subdifferential. It turns out that this can be done with the aid of the support function. If A is a set in the dual space, then the function $sA(x) = \sup_{y \in A} y \cdot x$ is the *support function* of A. *The necessary and sufficient condition for the equality* $s\partial p = p$ *is that* p *be a closed sublinear function.* Similarly, the equality $\partial sA = A$ holds if and only if A is a convex and closed set. These are duality relations as well, though unlike the case of involutive operators, as was, e.g., the conjugation operator for convex functions, they are generated by operators ∂, which map sublinear functions into convex sets, and s, which maps convex sets into sublinear functions.

There are a number of *operations* that transform convex objects again into convex ones. So are, e.g., the algebraic summation of sets or the usual summation of functions. According to the duality principle discussed above, each operation has its dual. For instance, consider the duality related to the Legendre–Young–Fenchel transform and the summation operation for functions. We can ask, which operation on conjugate functions gives rise to the conjugate function of the sum of two functions, or, in other words, how can we obtain $(f_1 + f_2)^*$ from f_1^* and f_2^*? This question can be answered with the aid of the operation

$$(f_1 \oplus f_2)(x) := \inf(f_1(x_1) + f_2(x_2) \mid x_1 + x_2 = x\},$$

which is called *the convolution* (or sometimes, *the infimal convolution*). Under some assumptions on the functions the formula

$$(f_1 + f_2)^* \;=\; f_1^* \oplus f_2^*$$

holds, and the equality

$$(f_1 \oplus f_2)^* = f_1^* + f_2^*$$

is fulfilled in any case. In this sense the operations $+$ and $\oplus$ are dual to each other. (Recently, the duality between linear operations and convolution caused the appearance of a new type of analysis parallel to the linear analysis (the so-called *idempotent analysis*).)

Formulas similar to those presented above constitute convex calculus. There are more than 30 of them (see Section 3). The application of the duality operators listed above results sometimes in unusual operations on sets or functions (e.g., the so-called Kelley sum for sets and functions).

The fundamentals of convex calculus are exposed in Sections 1 through 3 of Chapter 1. In particular, the theorems mentioned above are proved there.

7. Finite-dimensional convex geometry.

Finite-dimensional geometry is the cradle of convex analysis. The theory of convex sets began from the theory of convex polyhedra. Elements of this theory were laid by Cauchy (though the problem of classification of regular or semi-regular polyhedra was posed and to a large extent solved in antiquity, in particular, by the Plato school, Archimedes, and others).

Cauchy proved a remarkable theorem: If there are two convex polyhedra with congruent respective faces, then they are congruent themselves. But this direction developed into a systematic theory only in the works of Minkowski. He turned from polyhedra to convex bodies. Among fundamental results obtained in the 19th and early 20th centuries, one has to list Minkowski's theorem on existence and uniqueness of a polyhedron with given normals and areas of corresponding faces, Steiner's formula for the volume of the sum of a polyhedron and Euclidean sphere, Steiner's symmetrization, the Brunn–Minkowski inequality, the Blaschke compactness theorem, and many others. All these results are presented in this book.

Recent progress in mathematics gave a new incentive to the development of the theory of convex polyhedra. One source of interest in this theory was provided by the studies on solvability of algebraic equations in multidimensional spaces and on partial differential equations. Moreover, numerous links between convexity and algebraic geometry were discovered.

8. Convex extremal problems.

In this book we treat traditional topics for the theory of extremal problems, viz., existence of solutions, conditions for extremum, and algorithms.

A notable feature of convex problems is that in nondegenerate situations (we will explain later on what it means precisely) necessary conditions for extremum are also sufficient.

Let $f_i\colon \mathbb{R}^n \to \mathbb{R} \cup \{+\infty\}$, $i = 0, 1, \ldots, m$, be convex functions and A a convex subset of $\mathbb{R}^n$. The problem

$$(\mathcal{P}) \qquad f_0(x) \to \min, \quad f_i(x) \le 0, \qquad 0 \le i \le m,\ x \in A,$$

is referred to as a *convex problem* or a *convex programming problem.*

If all functions in the problem $(\mathcal{P})$ are affine, then $(\mathcal{P})$ is called a *linear programming problem.*

If $(\mathcal{P})$ contains no constraints, i.e., the problem is

$$f(x) \to \min, \quad x \in \mathbb{R}^n$$

(which is called a *problem without constraints*), then the necessary and sufficient condition for f to attain the minimum at a point $\widehat{x}$ is inclusion $0 \in \partial f(\widehat{x})$. This inclusion, which can be verified trivially, is an analog of the Fermat theorem in calculus.

Finite-dimensional problems with constraints were considered by Lagrange in the 17th century. For the treatment of such problems he proposed a general principle, which is known now as the *Lagrange principle.* This principle says that for obtaining necessary conditions for a minimum in a problem with constraints, one has to compose a function (called a *Lagrange function*), which is the sum of the functional to be minimized and constraints with indeterminate factors (called *Lagrange multipliers*), and to "minimize this function as if all variables were independent." For the problems considered by Lagrange this formulation has to be stated more precisely, but for the problems of convex programming it is correct literally: *if $\widehat{x}$ solves the problem $(\mathcal{P})$, then there are nonnegative Lagrange multipliers $\lambda_0, \lambda_1, \dots, \lambda_m$, which are not all equal to zero, such that $\widehat{x}$ is a minimum point on A of the Lagrange function $\mathcal{L}(x, \lambda_0, \lambda_1, \dots, \lambda_m)$ (this relation is called the minimum principle) and the complementary slackness conditions $\lambda_i f_i(\widehat{x}) = 0$, $i = 1, \dots, m$, are fulfilled.*

This is the *Karush–Kuhn–Tucker theorem.* And if $\lambda_0 \neq 0$ (this is the nondegeneracy mentioned above), then the condition that the Lagrange multipliers are nonnegative, the complementary slackness conditions, and the minimum principle are sufficient for the element $\widehat{x}$ to solve the problem $(\mathcal{P})$. Thus in the problems of convex programming, necessary conditions "almost" coincide with sufficient ones.

Besides these two notable features of convex problems (the most complete form of the Lagrange principle and coincidence of necessary and sufficient conditions), there is one more feature, which is related to duality of convex analysis objects discussed above: every convex programming problem has its dual, and it is worthwhile to investigate both problems jointly.

Let us outline the construction of a dual problem. Let $f\colon \mathbb{R}^n \to \mathbb{R} \cup \{\pm\infty\}$. Consider the problem

$$(\mathcal{P}_0) \qquad f(x) \to \min, \quad x \in \mathbb{R}^n.$$

The problem $(\mathcal{P})$ can be reduced to this form by setting $f(x) = f_0(x)$ if x satisfies the constraints in $(\mathcal{P})$ and $f(x) = +\infty$ otherwise.

Let $F\colon \mathbb{R}^n \times \mathbb{R}^m \to \mathbb{R} \cup \{+\infty\}$. With each $y \in \mathbb{R}^m$ we associate the problem

$$(\mathcal{P}_y) \qquad F(x, y) \to \min, \quad x \in \mathbb{R}^n.$$

The family of such problems is the *perturbation of the problem* $(\mathcal{P}_0)$. The *dual problem* to $(\mathcal{P}_0)$ (with respect to the given perturbation) is the problem

$$(\mathcal{P}_0^*) \qquad -F^*(0, y') \to \max, \quad y' \in (\mathbb{R}^m)',$$

where $F^*\colon (\mathbb{R}^n)' \times (\mathbb{R}^m)' \to \mathbb{R} \cup \{+\infty\}$ is the function conjugate with F. This duality construction is based on the Fenchel–Moreau theorem.

A few words about the methods of convex optimization, i.e., the algorithms for solving the convex problems. We mention here the simplex method and the so-called method of sections.

The simplex method proposed by G. Dantzig is the method of efficient descent along the vertices of the polyhedron of the admissible elements in a linear programming problem.

The underlying idea of the method of sections may be demonstrated by the univariate problem of smooth convex programming. Suppose we look for the absolute minimum of a convex smooth function f defined on the interval $[0, 1]$, given that we can compute the derivative of f at any desired point. The question is, what amount of computation is needed in order to localize the minimum by an interval of length ε?

It is not hard to see that the optimal method for finding the minimum point is the "method of bisection": first we compute $f'(0)$ and $f'(1)$. If they are of the same sign, we take the minimal of them, and the value of f at this point is the minimum. Otherwise we divide the interval into two halves, compute the derivative at the mid-point, and cut off the half-interval where the derivatives at the end-points have the same sign. On the remaining interval we proceed in the same way, and so on. Eventually we will come arbitrarily close to the minimum point of f. An important theorem of finite-dimensional geometry, the Grünbaum–Hammer theorem, carries over this method to the multidimensional case.

Besides these two methods, the book presents the method of ellipsoids and some recent work on the construction of algorithms for solving convex extremal problems.

The theory of convex extremal problems is presented in Section 4 of Part 1.

9. Applications.

The second part of our book deals with applications of the convexity theory to the various areas of mathematics, as well as to economics, technology, and the natural sciences.

The mathematical applications can be briefly listed as follows:

- Duality between convex calculus of subspaces and the theory of linear equations (see Section 7 where the Fredholm alternative is discussed).
- Duality of convex calculus of cones and the theory of linear inequalities (see Section 7 where the Farkas, Gale, and Ky Fan theorems are discussed).
- Duality of convex calculus of zero-convex sets and finite-dimensional geometry (see H. Weyl's theorem in Section 7).
- Calculus of support functions, theory of recovery, and inequalities for derivatives (Sections 9 and 10).
- Calculus of Minkowski's functions and classical inequalities (Section 8).

Section 8 also contains applications of the convex optimization to geometry and technology.

Many applications of convexity to problems of analysis, economics, and technology rely on the fundamental relationship between convexity and integration. This can be illustrated by the theorem on vector measures due to A. A. Lyapunov. In the simplest setting it reads as follows. Let $p_1(\cdot), \dots, p_n(\cdot)$ be n integrable functions on the interval $[0, 1]$. With each measurable set $E \subset [0, 1]$ we associate the point $x(E) = (x_1(E), \dots, x_n(E)) \in \mathbb{R}^n$, where $x_i(E) = \int_E p_i(t)dt$, $1 \leq i \leq n$. Then the set of such points corresponding to all measurable subsets E of $[0, 1]$ is a convex compact set in $\mathbb{R}^n$. This phenomenon explains the role of convexity in the calculus of variations and optimal control (the fundamental results, such as Legendre and Weierstrass conditions, Pontryagin's maximum principle, and others, rely on convexity resulting from integration). Connections of convex analysis with general theory of extremum are presented in Section 13. Clearly, it is due to the optimization theory that convex analysis is so widely applicable in geometry, approximation theory, theory of inequalities, and so on. A typical situation in economics is when many small elements are summed up. Such sums are naturally approximated by integrals, which also explains the role of convex analysis in mathematical economics.

The third part of the book is the Appendix. Here we prove some theorems which are fundamental for the convexity theory and review some aspects of the modern development of convex analysis.

This is briefly the contents of the book.

CHAPTER 1

Theory

1. Basic definitions

- In this section we give the basic facts about the structure of the space $\mathbb{R}^n$, study the properties of convex sets in $\mathbb{R}^n$ and convex functions on $\mathbb{R}^n$, and prove the separation theorems.

1.1. The space $\mathbb{R}^n$.

In this chapter we study convex objects in the space $\mathbb{R}^n$ (with n a positive integer) which consists of all ordered n-tuples of real numbers. Each n-tuple will be represented by a column of these numbers, but to save space we will write it as $x = (x_1, \ldots, x_n)^T$, where T means the transpose. The n-tuples x will be called *vectors* (or *column vectors*). For the vectors in $\mathbb{R}^n$ the operations of their addition and multiplication by a number are defined in a natural way: if $x = (x_1, \ldots, x_n)^T$ and $y = (y_1, \ldots, y_n)^T$, then $x + y = (x_1 + y_1, \ldots, x_n + y_n)^T$, and if $x = (x_1, \ldots, x_n)^T \in \mathbb{R}^n$ and $\alpha \in \mathbb{R}$, then $\alpha x = (\alpha x_1, \ldots, \alpha x_n)^T$.

The space $\mathbb{R}^n$ has dimension n (written $\dim \mathbb{R}^n = n$) in the sense that there exist n linearly independent vectors $e_1, \ldots, e_n$ (i.e., the equality $\alpha_1 e_1 + \cdots + \alpha_n e_n = 0$ is possible only for $\alpha_1 = \ldots = \alpha_n = 0$) such that any vector $x \in \mathbb{R}^n$ can be represented as $x = \xi_1 e_1 + \cdots + \xi_n e_n$ and the (uniquely defined) numbers $\xi_1, \ldots, \xi_n$ are the coordinates of x in the basis $e_1, \ldots, e_n$. A *basis* of $\mathbb{R}^n$ is any collection of n linearly independent vectors in $\mathbb{R}^n$. Obviously, the vectors $e_1 = (1, 0, \ldots, 0)^T, \ldots, e_n = (0, \ldots, 0, 1)^T$ form a basis in $\mathbb{R}^n$. It is the *standard basis*, and the components of $x = (x_1, \ldots, x_n)^T$ are the coordinates of vector x in this basis.

Let $l\colon \mathbb{R}^n \to \mathbb{R}$ be a linear functional, i.e., $l(x + y) = l(x) + l(y)$ for any $x, y \in \mathbb{R}^n$ and $l(\alpha x) = \alpha l(x)$ for any $x \in \mathbb{R}^n$ and $\alpha \in \mathbb{R}$. Then if $e_1, \ldots, e_n$ is the standard basis in $\mathbb{R}^n$ and $l(e_i) = x'_i$, $i = 1, \ldots, n$, then for any $x \in \mathbb{R}^n$ we have $l(x) = \sum_{i=1}^n x'_i x_i$. This sum may be written as a (matrix) product of the row vector $x' = (x'_1, \ldots, x'_n)$

by the column vector $x = (x_1, \dots, x_n)^T$, which will be written as $x' \cdot x$. Thus $l(x) = x' \cdot x$ for any $x \in \mathbb{R}^n$. On the other hand, any row vector x' defines a linear functional $x \to x' \cdot x$ on $\mathbb{R}^n$. Therefore the set of all linear functionals on $\mathbb{R}^n$ (called the *dual space* of $\mathbb{R}^n$) may be identified with the space $\mathbb{R}^n$ itself, with elements written as row vectors. Nevertheless, in view of subsequent generalizations, it is expedient to distinguish between $\mathbb{R}^n$ and its dual (the set of all ordered n-tuples of real numbers written in a row with the same operations of addition and multiplication by a number), which will be denoted by $(\mathbb{R}^n)'$.

There are some important classes of subsets of $\mathbb{R}^n$ which will be frequently encountered. These are, first of all, subspaces, affine subspaces, hyperplanes, and half-spaces.

A subset $L \subset \mathbb{R}^n$ is a *subspace* if $x + y \in L$ for any $x, y \in L$ and $\alpha x \in L$ for any $x \in L$ and $\alpha \in \mathbb{R}$. In any subspace $L \subset \mathbb{R}^n$ one can find a set of linearly independent vectors $e_1, \dots, e_m$, $m \leq n$, such that any $x \in L$ is representable as $x = \xi_1 e_1 + \dots + \xi_m e_m$, and each such set has the same cardinality m, which is called the *dimension* of L, written $\dim L = m$. It is proved in linear algebra that any subspace in $\mathbb{R}^n$ may be specified as the set of solutions to a linear system of homogeneous equations $a^i \cdot x = 0$, $i = 1, \dots, s$, where $a^i \in (\mathbb{R}^n)'$. Subspaces in the plane are straight lines passing through the origin (one-dimensional subspaces), and in the three-dimensional space these are straight lines and planes passing through the origin (one- and two-dimensional subspaces respectively).

A subset $M \subset \mathbb{R}^n$ is an *affine subspace* or *affine manifold* if M is a translated subspace, i.e., a set of the form $x + L$, where L is a subspace. The dimension of L is the *dimension of* M. These are all straight lines in the plane, and all straight lines and planes in the three-dimensional space. Let $a = (a_1, \dots, a_n) \in (\mathbb{R}^n)'$. A *hyperplane* $H(a, \gamma)$ (determined by a vector $a \in (\mathbb{R}^n)'$ and a number γ) is the set $\{x \in \mathbb{R}^n \mid a \cdot x = \gamma\}$ (i.e., a level set of a linear functional). Clearly, a hyperplane is an affine subspace, and for $\gamma = 0$ it is a subspace. The set $H(a, \gamma)$ in the plane consists of the points satisfying the equation $a_1 x_1 + a_2 x_2 = \gamma$, which is a straight line, and $H(a, \gamma)$ in the three-dimensional space is a plane. A hyperplane divides the space $\mathbb{R}^n$ into two sets (with the hyperplane being their common boundary). We denote these sets by $H_+(a, \gamma) = \{x \in \mathbb{R}^n \mid a \cdot x \leq \gamma\}$ and $H_-(a, \gamma) = \{x \in \mathbb{R}^n \mid a \cdot x \geq \gamma\}$. These are half-planes in the plane and half-spaces in the three-dimensional space.

In the sequel an important role will be played by straight lines, rays, and line segments, i.e., one-dimensional subspaces and affine subspaces, one-dimensional cones and convex sets (to be precisely defined later). Let $x, y \in \mathbb{R}^n$ and $x \neq y$. A *line segment* $[x, y]$ is defined as the set of points of the form $z = (1-\alpha)x + \alpha y$, $0 \le \alpha \le 1$. A *ray* originating from x and passing through y is the set of points of the form $z = (1-\alpha)x + \alpha y$, $\alpha \ge 0$, and, finally, the *straight line* passing through the points x and y is the set $z = (1-\alpha)x + \alpha y$, $\alpha \in \mathbb{R}$. If $n = 1, 2$, or 3, these are the usual line segments, rays, and straight lines.

Next, we will describe the so-called topological structure of $\mathbb{R}^n$. With each $x \in \mathbb{R}^n$ we associate the number $|x| = \sqrt{x^T \cdot x} = \sqrt{\sum_{i=1}^n x_i^2}$, called the *length* or the *Euclidean norm* of x. This quantity has the following properties: 1) $|x| \ge 0$ with $|x| = 0$ if and only if $x = 0$; 2) $|\alpha x| = |\alpha||x|$ for any $x \in \mathbb{R}^n$ and $\alpha \in \mathbb{R}$; 3) $|x + y| \le |x| + |y|$ for any $x, y \in \mathbb{R}^n$. The first two properties are obvious, and the third is established with the aid of the Cauchy–Schwarz inequality $|x^T \cdot y| \le |x||y|$, which can be easily verified. The quantity $d(x, y) = |x - y|$ is the *distance* between the vectors x and y.

Now we define open and closed sets in $\mathbb{R}^n$. Let $x \in \mathbb{R}^n$ and $r > 0$. The set $\dot{B}_r(x) = \{\xi \in \mathbb{R}^n \mid |\xi - x| < r\}$ is the *open ball* with center x and radius r.

Let $A \subset \mathbb{R}^n$ and $x \in A$. Then x is said to be an *interior point* of A if x is contained in A together with some open ball centered at x. The set of interior points of A is the *interior* of A, written $\operatorname{int} A$. A set $G \subset \mathbb{R}^n$ is *open* if each of its points is an interior one, i.e., $\operatorname{int} G = G$. A *neighborhood* of $x \in \mathbb{R}^n$ is any open set containing x.

In order to get accustomed to the basic topological notions, we advise the reader to think over the following exercises.

EXERCISE 1. Prove that any open ball is an open set.

EXERCISE 2. Prove that the union of any family of open sets and the intersection of finitely many open sets is an open set.

EXERCISE 3. Prove that if G is an open set, then for any $y \in \mathbb{R}^n$ and $\alpha \neq 0$ the set $y + \alpha G = \{z \in \mathbb{R}^n \mid z = y + \alpha x,\ x \in G\}$ is also open.

A set $F \subset \mathbb{R}^n$ is *closed* if its complement $\mathbb{R}^n \setminus F$ is open.

EXERCISE 4. Prove that the ball (sometimes termed the closed ball) $B_r(x) = \{\xi \in \mathbb{R}^n \mid |\xi - x| \leq r\}$ of radius r with center at x is a closed set.

EXERCISE 5. Prove that the intersection of any family of closed sets and the union of finitely many closed sets is a closed set.

Let $A \subset \mathbb{R}^n$. A point $x \in \mathbb{R}^n$ is an *accumulation point* of A if any neighborhood of x contains at least one point of A. The set of all accumulation points of a set A is called the *closure* of A and is denoted by $\operatorname{cl} A$. It is clear that $A \subset \operatorname{cl} A$.

EXERCISE 6. Prove that $\operatorname{cl} A$ is a closed set and if $A \subset B$, then $\operatorname{cl} A \subset \operatorname{cl} B$.

EXERCISE 7. Prove that a set A is closed if and only if it coincides with its closure; i.e., $A = \operatorname{cl} A$.

A sequence $\{x_k\}_{k \in \mathbb{N}}$ of points in $\mathbb{R}^n$ is said to converge to a point $x \in \mathbb{R}^n$ if for any $\varepsilon > 0$ there exists a positive integer $N = N(\varepsilon)$ such that $|x_k - x| < \varepsilon$ for any $k \geq N$.

EXERCISE 8. Prove that x is an accumulation point of A if and only if there exists a sequence of points in A converging to x.

EXERCISE 9. Prove that any convergent sequence $\{x_k\}_{k \in \mathbb{N}}$ is bounded; i.e., there exists $c \geq 0$ such that $|x_k| \leq c$ for all $k \in \mathbb{N}$.

A set $A \subset \mathbb{R}^n$ is *compact* if any sequence of its elements contains a subsequence converging to an element of A.

EXERCISE 10. Prove that a set in $\mathbb{R}^n$ is compact if and only if it is bounded (i.e., is contained in some ball) and closed.

Let $A \subset \mathbb{R}^n$ and $x_0 \in A$. A function $f\colon A \to \mathbb{R}$ is *continuous at the point* $x_0 \in A$ if for any $\varepsilon > 0$ there exists $\delta = \delta(\varepsilon) > 0$ such that $|f(x) - f(x_0)| < \varepsilon$ for any $x \in A$ such that $|x - x_0| < \varepsilon$ (in other words, for any $x \in A \cap B_\varepsilon(x_0)$).

This definition of continuity of a function at a point is equivalent to the following one: a function $f\colon A \to \mathbb{R}$ is continuous at $x_0 \in A$ if for any sequence of vectors $\{x_k\}_{k \in \mathbb{N}} \subset A$ converging to x_0 the sequence $\{f(x_k)\}_{k \in \mathbb{N}}$ converges to $f(x_0)$.

If a function is continuous at each point of A, it is said to be continuous on A.

EXERCISE 11. Prove that a function $f\colon \mathbb{R}^n \to \mathbb{R}$ is continuous on $\mathbb{R}^n$ if and only if the inverse image of any open (closed) set is open (closed). (The inverse image of a set A under a mapping $F\colon X \to Y$ is the set $F^{-1}(A) = \{x \in X \mid F(x) \in A\}$.)

EXERCISE 12. Prove that the functions $x \to |x|$ and $x \to x' \cdot x + \alpha$, where $x' \in (\mathbb{R}^n)'$ and $\alpha \in \mathbb{R}$, are continuous.

Exercises 11 and 12 imply that hyperplanes and half-spaces in $\mathbb{R}^n$ are closed sets, hence so are subspaces and affine subspaces in $\mathbb{R}^n$.

THEOREM (Weierstrass). *A function continuous on a compact set in $\mathbb{R}^n$ (i.e., on a bounded closed set) attains its maximum and minimum values on this set.*

PROOF. Let A be a compact subset of $\mathbb{R}^n$ and $f\colon A \to \mathbb{R}$ a continuous function. Denote $\gamma = \sup\{f(x) \mid x \in A\}$ and let $\{x_k\}$ be a sequence of elements of A such that $\{f(x_k)\}$ converges to γ. Select from $\{x_k\}$ a subsequence $\{x_{k_l}\}$ converging to $\overline{x} \in A$. Then the sequence $\{f(x_{k_l})\}$ converges to $f(\overline{x}) = \gamma$; hence γ is finite and $\overline{x}$ is a point where f takes its maximal value. The reasoning for minimum is similar. □

REMARK. The notions of open and closed sets, accumulation points, continuity, and compactness are so-called topological notions. They are introduced in $\mathbb{R}^n$ in a natural way based actually only on the notion of an open ball. In an infinite-dimensional space these notions, in general, are defined differently. We will set out convex analysis in infinite-dimensional spaces in Section 6. Presently we mean $\mathbb{R}^n$ only when speaking of topological notions. But it is worthwhile to note that all the results presented in this chapter remain valid in the infinite-dimensional setting if we specify the precise meaning of the words "closedness" and "closure".

1.2. Convex sets and functions.

- Convex sets and functions are the main objects of convex analysis. Here we present the basic concepts related to these objects.

We begin with convex sets. Recall that for $x, y \in \mathbb{R}^n$, the set $[x, y] = \{z \in \mathbb{R}^n \mid z = (1-\alpha)x + \alpha y,\ 0 \le \alpha \le 1\}$ is referred to as the *line segment (joining the points x and y)*. In a similar way, we define the *half-intervals* $[x, y) = \{z \in \mathbb{R}^n \mid z = (1-\alpha)x + \alpha y,\ 0 \le \alpha < 1\}$, $(x, y] = \{z \in \mathbb{R}^n \mid z = (1-\alpha)x + \alpha y,\ 0 < \alpha \le 1\}$ and the *interval* $(x, y) = \{z \in \mathbb{R}^n \mid z = (1-\alpha)x + \alpha y,\ 0 < \alpha < 1\}$.

DEFINITION 1. A nonempty set $A \subset \mathbb{R}^n$ is *convex* if, along with any two of its points x and y, it contains the line segment $[x, y]$.

The empty set is convex by definition.

For any $a \in (\mathbb{R}^n)'$ and $\gamma \in \mathbb{R}$, the hyperplane $H(a, \gamma) = \{x \in \mathbb{R}^n \mid a \cdot x = \gamma\}$ and the corresponding half-spaces $H_+(a, \gamma) = \{x \in \mathbb{R}^n \mid a \cdot x \leq \gamma\}$ and $H_-(a, \gamma) = \{x \in \mathbb{R}^n \mid a \cdot x \geq \gamma\}$ are convex sets.

The basic operations on convex sets are *intersection* and *algebraic sum*:

(1) Let $\{A_i\}_{i \in \mathcal{J}}$ be an arbitrary family of convex subsets of $\mathbb{R}^n$. Then $\bigcap_{i \in \mathcal{J}} A_i$ is a convex set (i.e., the intersection of any family of convex sets is a convex set).

In particular, the intersection of any family of half-spaces is a convex set. Moreover, this set is closed (see Exercise 5 above). Later we will show that any convex closed set is the intersection of some family of half-spaces (which is one of the basic results of convex analysis).

(2) Let $\{A_i\}_{i=1}^n$ be a finite family of convex sets in $\mathbb{R}^n$. Then $A_1 + \cdots + A_n := \{x \in \mathbb{R}^n \mid x = x_1 + \cdots + x_n,\ x_i \in A_i,\ i = 1, \ldots, n\}$ is a convex set (i.e., the algebraic sum of finitely many convex sets is a convex set).

If A is a convex subset of $\mathbb{R}^n$ and $\lambda \in \mathbb{R}$, then the homothetic transformation of A with coefficient λ, i.e., the set $\lambda A := \{\lambda x \mid x \in A\}$, is a convex set. More generally, if $\Lambda \colon \mathbb{R}^n \to \mathbb{R}^m$ is a linear operator and B a convex subset of $\mathbb{R}^n$, then $\Lambda(A)$, the image of A under the mapping Λ (to be written also ΛA), and $\Lambda^{-1}(B)$, the inverse image of B under the mapping Λ (to be written also $B\Lambda$), are convex sets. These assertions are deduced directly from the definitions.

A nonempty set $K \subset \mathbb{R}^n$ is a *cone* if, along with any element $x \in K$, it contains the element $\alpha x \in K$ for any $\alpha > 0$.

It can be easily verified that a cone K is convex if and only if $K + K \subset K$.

The set $\mathbb{R}^n_+ = \{x = (x_1, \ldots, x_n) \in \mathbb{R}^n \mid x_i \geq 0,\ i = 1, \ldots, n\}$ is a convex cone.

Let $x_i \in \mathbb{R}^n$ and $\alpha_i \in \mathbb{R}$, $i = 1, \ldots, n$. The vector $x = \sum_{i=1}^n \alpha_i x_i$ is called

(a) a *linear combination* of the vectors $x_1, \ldots, x_n$ for any possible values $\alpha_i \in \mathbb{R}$, $i = 1, \ldots, n$. A set containing any linear combination of any two of its elements (i.e., with any two of its points containing the plane passing through them and the origin) is a *subspace* (this notion, as well as some other notions given below, has appeared before, but here we define them in a unified manner);

(b) an *affine combination* of $x_1, \dots, x_n$ if $\alpha_i \in \mathbb{R}$, $i = 1, \dots, n$, and $\sum_{i=1}^n \alpha_i = 1$. A set containing any affine combination of any two of its elements (i.e., with any two of its points containing the straight line passing through them) is an *affine subspace* or *affine manifold*;

(c) a *conic combination* of $x_1, \dots, x_n$ if $\alpha_i \geq 0$, $i = 1, \dots, n$. A set containing any conic combination of any two of its elements (i.e., with any two of its points containing the planar angle between the rays determined by these points) is a *convex cone*;

(d) a *convex combination* of $x_1, \dots, x_n$ if $\alpha_i \geq 0$, $i = 1, \dots, n$, and $\sum_{i=1}^n \alpha_i = 1$. A set containing any convex combination of any two of its elements (i.e., with any two of its points containing the line segment joining them) is a *convex set.*

If A is a nonempty subset of $\mathbb{R}^n$, then we immediately deduce from the definitions that the set of

(a) all linear combinations of elements of A is the least vector subspace in $\mathbb{R}^n$ containing A. It is called the *linear span* of A and is denoted by $\operatorname{span} A$;

(b) all affine combinations of elements of A comprise the least affine subspace in $\mathbb{R}^n$ containing A, i.e., the set of the form $x + L_A$, where $x \in A$ and L_A is a subspace in $\mathbb{R}^n$. It is called the *affine hull* of A and is denoted by $\operatorname{aff} A$;

(c) all conic combinations of elements of A comprise the least convex cone in $\mathbb{R}^n$ containing A. It is called the *conic hull* of A and is denoted by $\operatorname{cone} A$;

(d) all convex combinations of elements of A comprise the least convex set in $\mathbb{R}^n$ containing A. It is called the *convex hull* of A and is denoted by $\operatorname{co} A$.

If A is a convex subset of $\mathbb{R}^n$ and $\operatorname{aff} A = x + L_A$, then the *dimension of* A (written $\dim A$) is the dimension of L_A.

The points $a_1 \dots, a_{m+1} \in \mathbb{R}^n$ $(m \geq 1)$ are *affine independent* if $\sum_{i=1}^{m+1} \lambda_i a_i = 0$ and $\sum_{i=1}^{m+1} \lambda_i = 0$ entail $\lambda_1 \cdots = \lambda_{m+1} = 0$. It can be easily verified that the points $a_1, \dots, a_{m+1}$ are affine independent if and only if for any $1 \leq i_0 \leq m+1$ the vectors $a_i - a_{i_0}$, $1 \leq i \leq m+1$, $i \neq i_0$, are linearly independent.

The convex hull of affine independent points $a_1, \dots, a_{m+1}$ is referred to as an *m-dimensional simplex*, and the points $a_1, \dots, a_{m+1}$ as its *vertices*. Any point x of the simplex is uniquely representable as $x = \sum_{i=1}^{m+1} \lambda_i a_i$, where $\lambda_i \geq 0$, $i = 1, \dots, m+1$, and $\sum_{i=1}^{m+1} \lambda_i = 1$. The coefficients λ_i, $1 \leq i \leq m+1$, are the *barycentric coordinates* of x. It is clear that a one-dimensional simplex is a line segment, a two-dimensional simplex is a triangle, and a three-dimensional simplex is a tetrahedron.

Let A be a convex subset of $\mathbb{R}^n$. If its dimension is greater than zero, then it equals the maximal dimension of simplexes contained in A. Indeed, let $\dim A = m$. It follows from the definition of the dimension of a convex set that the maximal dimension of simplexes contained in A is no greater than m. Assume that it is less than m, and let $a_1, \dots, a_{k+1}$ $(k < m)$ be the maximal affine independent set of points in A. If $A \subset \operatorname{aff}\{a_1, \dots, a_{k+1}\}$, then $\operatorname{aff} A \subset \operatorname{aff}\{a_1, \dots, a_{k+1}\}$ in contradiction with $\dim A = m$. Let $b \in A \setminus \operatorname{aff}\{a_1, \dots, a_{k+1}\}$. But in this case one can easily verify that the points $b, a_1, \dots, a_{k+1}$ are affine independent, which contradicts the maximality property of the set of a's.

It is clear that such sets as a line segment in the plane or a triangle in the three-dimensional space have no interior points. However, they have interior points relative to their affine hulls, which motivates introducing the following notion. Let A be a convex subset of $\mathbb{R}^n$. A point $x \in A$ is a *relatively interior* point if it has a neighborhood U such that $U \cap \operatorname{aff} A \subset A$. The set of all relatively interior points of A is the *relative interior* of A, written $\operatorname{ri} A$.

PROPOSITION 1. *Let A be a nonempty convex subset of $\mathbb{R}^n$. Then:*

(a) $\operatorname{ri} A \neq \emptyset$;

(b) *if $x_1 \in \operatorname{ri} A$ and $x_2 \in \operatorname{cl} A$, then all the points of the interval (x_1, x_2) belong to* $\operatorname{ri} A$;

(c) $\operatorname{ri} A$ *and* $\operatorname{cl} A$ *are convex sets and* $\operatorname{cl}(\operatorname{ri} A) = \operatorname{cl} A$.

PROOF. (a) Assume first that $\operatorname{aff} A = \mathbb{R}^n$, i.e., $\dim A = n$, so that A contains an n-dimensional simplex S. Let $a_1, \dots, a_{n+1}$ be its vertices. Without loss of generality we may assume that $a_1 = 0$. We will show that any point of S with positive barycentric coordinates belongs to $\operatorname{int} S$. Indeed, let $x_0 = \sum_{k=1}^{n+1} \lambda_k a_k$, $\sum_{k=1}^{n+1} \lambda_k = 1$, $\lambda_k > 0$, $k = 1, \dots, n+1$. If $e_1, \dots, e_n$ is the standard basis in $\mathbb{R}^n$ and $x = (x_1, \dots, x_n)^T \in \mathbb{R}^n$, then $x = \sum_{i=1}^{n} x_i e_i$. Further, since the vectors $a_2, \dots, a_{n+1}$ are linearly independent, we can write $e_i = \sum_{k=2}^{n+1} \alpha_{ik} a_k$, $1 \leq i \leq n$; hence $x_0 + x = \sum_{k=2}^{n+1} (\lambda_k + \sum_{i=1}^{n} x_i \alpha_{ik}) a_k = \sum_{k=2}^{n+1} \gamma_k a_k$. Since $\lambda_k > 0$, $2 \leq k \leq n+1$, and $\sum_{k=2}^{n+1} \lambda_k < 1$, there exists an $\varepsilon > 0$

such that $\gamma_k \geq 0$, $2 \leq k \leq n+1$, and $\sum_{k=2}^{n+1} \gamma_k \leq 1$ whenever $|x_i| < \varepsilon$, $1 \leq i \leq n$. Therefore $x_0 + x \in S$ for any x with such components and *a fortiori* for any vector in the open ball $\dot{B}_\varepsilon(0)$, i.e., $x_0 \in \operatorname{int} S = \operatorname{ri} S$.

If $\dim A = m < n$, we may identify $\operatorname{aff} A$ with the space $\mathbb{R}^m$, thus getting that $\operatorname{ri} A = \operatorname{int} A$ (in $\mathbb{R}^m$). Then the preceding arguments show that in this case $\operatorname{ri} A \neq \emptyset$ as well.

(b) In view of this remark we may assume that $\operatorname{aff} A = \mathbb{R}^n$ and hence $\operatorname{ri} A = \operatorname{int} A$. Let $y \in (x_1, x_2)$; then $y = (1-\alpha)x_1 + \alpha x_2$ for some $0 < \alpha < 1$. Since $x_2 \in \operatorname{cl} A$, we have $x_2 \in A + \dot{B}_\varepsilon(0)$ for any $\varepsilon > 0$. Then $y + \dot{B}_\varepsilon(0) = (1-\alpha)x_1 + \alpha x_2 + \dot{B}_\varepsilon(0) \subset (1-\alpha)x_1 + \alpha(A + \dot{B}_\varepsilon(0)) + \dot{B}_\varepsilon(0) = (1-\alpha)(x_1 + \dot{B}_\gamma(0)) + \alpha A$, where $\gamma = \varepsilon(1+\alpha)(1-\alpha)^{-1}$. By assumption, $x_1 \in \operatorname{int} A$ $x_1 + \dot{B}_\gamma(0) \subset A$ for ε small enough; therefore $y + \dot{B}_\varepsilon(0) \subset (1-\alpha)A + \alpha A = A$.

(c) We again assume that $\operatorname{aff} A = \mathbb{R}^n$. Convexity of $\operatorname{int} A$ follows directly from (b). Let us prove convexity of $\operatorname{cl} A$. Let $x_i \in \operatorname{cl} A$, $i = 1, 2$, and let $x = (1-\alpha)x_1 + \alpha x_2$, $0 \leq \alpha \leq 1$. For any $\varepsilon > 0$ we have $(x_i + \dot{B}_\varepsilon(0)) \cap A \neq \emptyset$, $i = 1, 2$. Let $\overline{x}_i \in (x_i + \dot{B}_\varepsilon(0)) \cap A$, $i = 1, 2$. Set $\overline{x} = (1-\alpha)\overline{x}_1 + \alpha \overline{x}_2$. Then $\overline{x} \in A$ and $\overline{x} \in (1-\alpha)(x_1 + \dot{B}_\varepsilon(0)) + \alpha(x_2 + \dot{B}_\varepsilon(0)) = (1-\alpha)x_1 + \alpha x_2 + \dot{B}_\varepsilon(0)$; i.e., any neighborhood of the point $x = (1-\alpha)x_1 + \alpha x_2$ has a nonempty intersection with A, so that $x \in \operatorname{cl} A$.

The inclusion $\operatorname{cl}(\operatorname{ri} A) \subset \operatorname{cl} A$ is obvious. Let $x \in \operatorname{cl} A$. By (a) there exists an $x_1 \in \operatorname{ri} A$. Then $(x, x_1) \subset \operatorname{ri} A$ by (b), which implies that $x \in \operatorname{cl}(\operatorname{ri} A)$. □

Now we turn to convex functions. We will have to deal with functions taking not only finite values but also $-\infty$ and $+\infty$. To this end we introduce the set $\overline{\mathbb{R}} = \mathbb{R} \cup \{-\infty\} \cup \{+\infty\}$, i.e., the real line completed with symbols $\pm\infty$, with order relation extended in a natural way, $-\infty < a < +\infty$ for $a \in \mathbb{R}$. Moreover, we assume that $a \pm \infty = \pm\infty$ for any $a \in \mathbb{R}$, $a(\pm\infty) = \pm\infty$ if $a > 0$, and $+\infty + (+\infty) = +\infty$, $-\infty + (-\infty) = -\infty$.

With each function $f \colon \mathbb{R}^n \to \overline{\mathbb{R}}$ we associate the set

$$\operatorname{dom} f := \{\, x \in \mathbb{R}^n \mid f(x) < +\infty \,\},$$

to be called the *effective set* of f.

The set

$$\operatorname{epi} f := \{(x, \alpha) \in \mathbb{R}^n \times \mathbb{R} \mid x \in \operatorname{dom} f,\ \alpha \geq f(x)\}$$

(where $\mathbb{R}^n \times \mathbb{R}$ is the space $\mathbb{R}^{n+1}$ regarded as the set of pairs (x, α), $x \in \mathbb{R}^n$, $\alpha \in \mathbb{R}$) is the *epigraph* of f.

A function $f\colon \mathbb{R}^n \to \overline{\mathbb{R}}$ is *proper* if at no point it takes the value $-\infty$ and is not equal to $+\infty$ identically.

DEFINITION 2. A function $f\colon \mathbb{R}^n \to \overline{\mathbb{R}}$ is *convex* if $\operatorname{epi} f$ is a convex set in $\mathbb{R}^n \times \mathbb{R}$.

For proper functions there is an equivalent definition of convexity which often proves to be more convenient. Namely, the following assertion holds: *a proper function $f\colon \mathbb{R}^n \to \mathbb{R} \cup \{+\infty\}$ is convex if and only if for any points $x_1, x_2 \in \mathbb{R}^n$ the graph of the function on the segment $[x_1, x_2]$ never lies above the line segment joining the points $f(x_1)$ and $f(x_2)$; i.e., for any $x_1, x_2 \in \mathbb{R}^n$ and any $\alpha_i \geq 0$, $i = 1, 2$, such that $\alpha_1 + \alpha_2 = 1$, the following inequality holds*:

$$f(\alpha_1 x_1 + \alpha_2 x_2) \leq \alpha_1 f(x_1) + \alpha_2 f(x_2),$$

which is known as the *Jensen inequality*.

Its proof follows easily from definitions.

Here are examples of convex functions on the real line: the quadratic function $x \to ax^2 + bx + c$ with $a \geq 0$; the exponential function $x \to e^{\alpha x}$, $\alpha \in \mathbb{R}$; powers of the absolute value, $x \to |x|^p$, $p \geq 1$; the function $x \to -\log x$ for $x > 0$ and $+\infty$ for $x \leq 0$; and the function $x \to -(x \log_2 x + (1 - x) \log_2(1 - x))$ for $0 < x < 1$, 0 for $x = 0$ or 1, and $+\infty$ otherwise (the last function is, up to sign, the entropy of the random variable taking the values 0 and 1 with probabilities x and $1 - x$).

An important (though uncustomary) convex function is the *indicator function* of a convex set A. It is defined as

$$\delta A(x) = \begin{cases} 0, & x \in A, \\ +\infty, & x \notin A. \end{cases}$$

For many functions (in particular, for most of those given above) it is convenient to establish their convexity using the following assertion: *if a function f is twice differentiable on (a, b) $(-\infty \leq a < b \leq +\infty)$ and $f''(x) \geq 0$ for any $x \in (a, b)$, then $f\colon \mathbb{R} \to \overline{\mathbb{R}}$, where $f(x) = +\infty$ for $x \notin (a, b)$, is a convex function.*

Indeed, since $f''(x) \geq 0$ for $x \in (a, b)$, we see that f' is nondecreasing. Therefore, if $a < x_1 < x_2 < b$, $0 < \alpha < 1$ and $x = (1 - \alpha)x_1 + \alpha x_2$, we have by the mean value theorem $f(x) - f(x_1) \leq f'(x)(x - x_1)$ and $f(x_2) - f(x) \geq f'(x)(x_2 - x)$, or $f(x) \leq f(x_1) +$

$\alpha f'(x)(x_2-x_1)$ and $f(x) \le f(x_2)-(1-\alpha)f'(x)(x_2-x_1)$. Multiplying the next-to-last inequality by $1-\alpha$ and the last one by α and adding them, we obtain $f((1-\alpha)x_1+\alpha x_2) \le (1-\alpha)f(x_1)+\alpha f(x_2)$. □

The main operations on convex functions are *maximum* and *summation*:

(a) Let $f_i\colon \mathbb{R}^n \to \overline{\mathbb{R}}$, $i = 1, 2$, be convex functions. Then $f_1 \vee f_2(x) = \max(f_1(x), f_2(x))$ is also a convex function. Actually, the following more general assertion is valid: if $f_j\colon \mathbb{R}^n \to \overline{\mathbb{R}}$, $j \in J$, is a family of convex functions, then $f = \sup_{j\in J} f_j$ is also a convex function and $\operatorname{epi} f = \bigcap_{j\in J} \operatorname{epi} f_j$ (i.e., the supremum of any family of convex functions is again a convex function).

(b) If $f, g\colon \mathbb{R}^n \to \overline{\mathbb{R}}$ are proper convex functions, then $f+g$ (i.e., $(f+g)(x) = f(x)+g(x)$) is also convex (i.e., the sum of convex functions is again a convex function).

Note also that, for a convex f and $\alpha > 0$, the product αf is a convex function.

Property (a) follows easily from the obvious fact that $\operatorname{epi} f = \bigcap_{j\in J} \operatorname{epi} f_j$ and the fact that the intersection of any family of convex sets is again a convex set. Property (b) and the statement about the product of a convex function by a positive number are easily verified with the aid of the Jensen inequality.

If $\Lambda\colon \mathbb{R}^n \to \mathbb{R}^m$ is a linear operator and $f\colon \mathbb{R}^m \to \overline{\mathbb{R}}$ and $g\colon \mathbb{R}^n \to \overline{\mathbb{R}}$ are convex functions, then it is not hard to see that the functions $f\Lambda\colon \mathbb{R}^n \to \overline{\mathbb{R}}$, $f\Lambda(x) = f(\Lambda x)$, called the inverse image of f under the mapping Λ, and $\Lambda g\colon \mathbb{R}^m \to \overline{\mathbb{R}}$, $\Lambda g(y) = \inf\{g(x) \mid \Lambda x = y\}$, called the image of g under the mapping Λ, are also convex.

An important class of convex functions consists of sublinear functions. A function $p\colon \mathbb{R}^n \to \overline{\mathbb{R}}$ is *sublinear* if $\operatorname{epi} p$ is a convex cone in $\mathbb{R}^n \times \mathbb{R}$.

The Jensen inequality implies that a proper function $p\colon \mathbb{R}^n \to \overline{\mathbb{R}}$ is sublinear if and only if

$$p(x+y) \le p(x)+p(y), \quad \forall x, y \in \mathbb{R}^n, \tag{p_1}$$

$$p(\alpha x) = \alpha p(x), \quad \forall x \in \mathbb{R}^n, \forall \alpha > 0 \tag{p_2}$$

(i.e., a proper sublinear function is a convex function positively homogeneous of first order).

For example, linear functions and the supremum of any family of linear functions are sublinear functions. (One of the important facts of convex analysis to be proved below is that a sublinear function with closed epigraph is the supremum of some family of linear functions).

Now we consider three important examples of sublinear functions: the Minkowski function, the support function, and the directional derivative of a convex function.

Let A be a nonempty subset of $\mathbb{R}^n$. The function $\mu A\colon \mathbb{R}^n \to \overline{\mathbb{R}}$ defined by the equality $\mu A(x) = \inf\{\lambda > 0 \mid x \in \lambda A\}$ is the *Minkowski function* of the set A.

Let A be a nonempty subset of $\mathbb{R}^n$. The function sA on $(\mathbb{R}^n)'$ defined by the equality $sA(x') = \sup_{x \in A} x' \cdot x$ is the *support function* of A. It is obvious that this function is positively homogeneous, and it is convex because $sA(\cdot)$ is a supremum of convex functions.

Now we define the directional derivative. Let $x_0, x \in \mathbb{R}^n$ and let the function $f\colon \mathbb{R}^n \to \overline{\mathbb{R}}$ be finite at x_0. The quantity

$$f'(x_0; x) = \lim_{t \downarrow 0} \frac{f(x_0 + tx) - f(x_0)}{t}$$

(whenever it exists) is the *derivative of f at the point x_0 in direction x*.

For convex functions the limit (possibly infinite) in the right-hand side of the above relation always exists. Indeed, the following assertion is valid: *if $f\colon \mathbb{R}^n \to \overline{\mathbb{R}}$ is a proper convex function and $x_0 \in \operatorname{dom} f$, then $f'(x_0; \cdot)$ is a sublinear function, and for any $x \in \mathbb{R}^n$*

$$f'(x_0; x) = \inf_{t > 0} \frac{f(x_0 + tx) - f(x_0)}{t}.$$

To prove this assertion, we will show first that for any $x \in \mathbb{R}^n$ the function $h\colon t \to t^{-1}\left(f(x_0 + tx) - f(x_0)\right)$, $t > 0$, is nondecreasing. Indeed, let $0 < s \le t$. Since $x_0 + sx = (s/t)(x_0 + tx) + ((t - s)/t)x_0$, we have by convexity of f

$$f(x_0 + sx) - f(x_0) \le \frac{s}{t}\left(f(x_0 + tx) - f(x_0)\right) + \frac{t - s}{t}\left(f(x_0) - f(x_0)\right),$$

i.e., $h(s) \le h(t)$. Thus the limit in the definition of the derivative can be replaced by infimum (which may be equal to $-\infty$).

Obviously, $f(x_0;\cdot)$ is homogeneous. Its convexity follows by a simple calculation:

$$\begin{aligned} f'(x_0;x_1+x_2) &= \lim_{t\downarrow 0} \frac{f(x_0+(t/2)(x_1+x_2))-f(x_0)}{t/2} \\ &\le \lim_{t\downarrow 0} \frac{f(x_0+tx_1)-f(x_0)+f(x_0+tx_2)-f(x_0)}{t} \\ &= f'(x_0;x_1)+f'(x_0;x_2). \end{aligned}$$

The main results of convex analysis (see Sections 2, 3 below) are based on separation theorems for convex sets.

Separation theorems in a finite-dimensional space. Recall that for $x' \in (\mathbb{R}^n)'$, $x' \neq 0$, and $\gamma \in \mathbb{R}$, the set $H = H(x',\gamma) = \{x \in \mathbb{R}^n \mid x'\cdot x = \gamma\}$ is a *hyperplane* in $\mathbb{R}^n$. It is clear that $H(x',\gamma) = H(cx', c\gamma)$ for any $c \neq 0$. The hyperplane $H(x',\gamma)$ determines two half-spaces $H_+(x',\gamma) = \{x \in \mathbb{R}^n \mid x'\cdot x \le \gamma\}$ and $H_-(x',\gamma) = \{x \in \mathbb{R}^n \mid x' \cdot x \ge \gamma\}$.

Two nonempty subsets, A and B, of $\mathbb{R}^n$ are said to be *separable* if there exists a hyperplane H such that A and B lie in the opposite half-spaces determined by H.

This geometric definition of separability is, obviously, equivalent to the following algebraic one: the sets A and B are separable if there exists a nonzero element $x' \in (\mathbb{R}^n)'$ such that

$$\sup_{a\in A} x'\cdot a \le \inf_{b\in B} x'\cdot b.$$

If the inequality is strict, then the sets A and B are *strictly separable* (in geometric terms it means that A and B can be separated by two different parallel hyperplanes).

We will prove first the theorem on strict separability of a point from a set, and then deduce from it the theorem on separability of sets.

THEOREM (second separability theorem or the theorem on strict separability of a point from a set). *Let A be a nonempty convex closed subset of $\mathbb{R}^n$ and let $b \notin A$. Then set A and point b are strictly separable.*

PROOF. Let $\overline{x} \in A$, $r = |\overline{x} - b|$, and $A_1 = A \cap B_r(b) = \{x \in \mathbb{R}^n \mid |x-b| \le r\}$. Consider the function $f\colon \mathbb{R}^n \to \mathbb{R}$, $f(x) = |x-b|$. This function is continuous; hence by the Weierstrass theorem it attains its infimum on the bounded closed set A_1 at some point $\widehat{x} \in A_1$. In other words, $\widehat{x}$ is the closest point to b in A_1, hence also in A. Set $x' = (\widehat{x} - b)^T$ and consider the hyperplane $H = \{x \in \mathbb{R}^n \mid x' \cdot x = x' \cdot \widehat{x}\}$. Then $x' \cdot b = x' \cdot (b - \widehat{x} + \widehat{x}) = -|x'|^2 + x' \cdot \widehat{x} < x' \cdot \widehat{x}$. It remains to show that $x' \cdot x \geq x' \cdot \widehat{x}$ for any $x \in A$. Suppose there exists $x_0 \in A$ such that $x' \cdot x_0 < x' \cdot \widehat{x}$. Since A is convex, $(1-t)\widehat{x} + tx_0 \in A$ for $0 \leq t \leq 1$, and we have

$$\begin{aligned} |(1-t)\widehat{x} + tx_0 - b|^2 &= |x' + t(x_0 - \widehat{x})|^2 \\ &= |x'|^2 + 2t\, x' \cdot (x_0 - \widehat{x}) + t^2 |x_0 - \widehat{x}|^2 \\ &< |x'|^2 = |\widehat{x} - b|^2 \end{aligned}$$

for small t. But this contradicts the property that $\widehat{x}$ is the closest point to b in A, hence $x' \cdot b < x' \cdot \widehat{x} \leq x' \cdot x$ for any $x \in A$; i.e., A and b are strictly separable. □

THEOREM (first separability theorem or the theorem on separability of sets). *Let A and B be nonempty convex subsets of $\mathbb{R}^n$ and let* $\operatorname{ri} A \cap B = \emptyset$. *Then A and B are separable.*

PROOF. Let $C = \operatorname{ri} A - B$. Since $\operatorname{ri} A$ is a convex set (see Proposition 1), the set C is convex and $0 \notin C$. Suppose $0 \notin \operatorname{cl} C$. Since $\operatorname{cl} C$ is also convex (Proposition 1), by the previous theorem there is a nonzero vector $x' \in (\mathbb{R}^n)'$ such that $x' \cdot x < 0$ for any $x \in \operatorname{cl} C$ and *a fortiori* for $x = a - b \in \operatorname{ri} A - B = C$, i.e., $x' \cdot a < x' \cdot b$ for any $a \in \operatorname{ri} A$ and $b \in B$. This implies that $\operatorname{ri} A$ belongs to the closed subspace $H_+(x', \gamma) = \{x \in \mathbb{R}^n \mid x' \cdot x \leq \gamma\}$, where $\gamma = \inf\{x' \cdot b \mid b \in B\}$ and $B \subset H_-(x', \gamma)$. But, by Proposition 1, $A \subset \operatorname{cl} A = \operatorname{cl}(\operatorname{ri} A) \subset H_+(x', \gamma)$; hence the sets A and B are separable.

Now let $0 \in \operatorname{cl} C$ and let $x \in \operatorname{ri} C$ ($\operatorname{ri} C \neq \emptyset$ by Proposition 1). Then for any $\lambda > 0$ the vector $-\lambda x$ does not belong to $\operatorname{cl} C$, since otherwise (again by Proposition 1) $0 = \lambda(1+\lambda)^{-1}x + (1+\lambda)^{-1}(-\lambda x)$ would belong to $\operatorname{ri} C$ and hence to C. Then there exists a sequence of points $\{x_k\}$ outside $\operatorname{cl} C$ such that $x_k \to 0$ as $k \to \infty$. According to the second separability theorem proved above, there are vectors $x'_k \in (\mathbb{R}^n)'$, $x'_k \neq 0$, $k \in \mathbb{N}$, such that

$$x'_k \cdot x < x'_k \cdot x_k \tag{i}$$

for each $k \in \mathbb{N}$ and $x \in \operatorname{cl} C$. Dividing (i) by $|x'_k|$, we can assume that $|x'_k| = 1$. Then (since the sphere $S = \{x \in \mathbb{R}^n \mid |x| = 1\}$ is bounded and closed) the sequence $\{x'_k\}$ contains a subsequence converging to some vector x'. Clearly $|x'| = 1$; hence $x' \neq 0$. The passage to a limit in inequality (i) along this subsequence yields $x' \cdot x \leq 0$. Then exactly the same arguments as in the preceding theorem show that the sets A and B are separable. □

2. Duality in convex analysis

- An important feature of convex objects is the possibility of their three-way description: namely, geometric, as a set of points, segments, rays, or straight lines with certain properties; analytic, as a superposition of some operators defined on families of convex sets and functions; and algebro-analytic, as sets of solutions of a family of equations or inequalities (for sets) or suprema of families of affine or linear functions (for functions).

In this section we formulate a theorem on convex duality which gives all such descriptions for the main convex objects.

We begin with a description of operators acting on classes of convex sets and functions.

A) *Conjugation operator for functions.* Let a function $f\colon \mathbb{R}^n \to \overline{\mathbb{R}}$ be given. The function $f^*\colon (\mathbb{R}^n)' \to \overline{\mathbb{R}}$ defined by the equality

$$f^*(x') := \sup_{x \in \mathbb{R}^n} (x' \cdot x - f(x))$$

is called the *conjugate function* of f, or the *Young* or *Young–Fenchel* or *Legendre* transform of f.

Since f^* is the supremum of convex closed functions $x' \to x' \cdot x - f(x)$, $x \in \mathbb{R}^n$, its epigraph f^* is the intersection of the epigraphs of these functions; hence f^* is a convex closed function on $(\mathbb{R}^n)'$. Thus, dealing only with convex functions, we have an operator *, which maps convex functions on $\mathbb{R}^n$ to convex closed functions on $(\mathbb{R}^n)'$.

Here are some examples of conjugate functions. 1) Let $n = 1$ and $f(x) = e^x$. Then $f^*(x') = x' \log x' - x'$ for $x' > 0$, $f^*(0) = 0$, and $f^*(x') = +\infty$ for $x' < 0$.

2) Let $f(x) = \frac{1}{p}\left(\sum_{i=1}^n |x_i|^p\right)$, where $x = (x_1, \dots, x_n)^T$ and $1 < p < \infty$. Then $f^*(x') = \frac{1}{p'}\left(\sum_{i=1}^n |x_i'|^{p'}\right)$, where $x' = (x_1', \dots, x_n')$ and $1/p + 1/p' = 1$.

3) Let A be a nonempty subset of $\mathbb{R}^n$. Then $(\delta A)^*(x') = sA(x')$ (the function conjugate with an indicator function of a set is the support function of this set).

4) Let $\widehat{x}' \in (\mathbb{R}^n)'$, $\alpha \in \mathbb{R}$, and let $f\colon x \to \widehat{x}' \cdot x - \alpha$ be an affine function. Then $f^*(x') = \alpha$ for $x' = \widehat{x}'$ and $f^*(x') = \infty$ otherwise. Such a function, with the epigraph consisting of a single ray, will be called elementary. The epigraph of any function is the union of epigraphs of elementary functions.

Note that if f is a convex proper closed function, then so is f^*. Indeed, its convexity and closedness follow from the definition. Since f is not identically $+\infty$, f^* nowhere equals $-\infty$. It will be shown in the duality theorem presented below that there exists an affine function $p(x) = x' \cdot x - \alpha$ which does not exceed f. Then $f^*(x') \geq p^*(x') = \alpha$, so that f^* is a proper function.

EXERCISE 1. Find the functions conjugate with $f(x_1, x_2) = ax_1 + bx_2 + c$, $f(x_1, \dots, x_n) = \max(x_1, \dots, x_n)$.

The *second conjugate* of f is the function $f^{**}\colon \mathbb{R}^n \to \overline{\mathbb{R}}$ defined by the equality

$$f^{**}(x) := \sup_{x' \in (\mathbb{R}^n)'} (x' \cdot x - f^*(x')).$$

By the same reasons as above, f^{**} is a convex closed function on $\mathbb{R}^n$.

EXERCISE 2. Find the second conjugate functions of $f(x) = \sin x$, $f(x) = 1/x^2$.

It follows directly from the definitions that for any $x \in \mathbb{R}^n$ and $x' \in (\mathbb{R}^n)'$ the following inequalities hold:

$$x' \cdot x \leq f(x) + f^*(x'), \quad x' \cdot x \leq f^*(x') + f^{**}(x),$$

which are known as the *Young inequalities*; moreover, it follows that $f^{**}(x) \leq f(x)$ for any $x \in \mathbb{R}^n$ and that $f(x) \leq g(x)$ for all $x \in \mathbb{R}^n$ entails $f^*(x') \geq g^*(x')$ for all $x' \in (\mathbb{R}^n)'$.

B) *Subdifferential.* Let $p\colon \mathbb{R}^n \to \overline{\mathbb{R}}$ be a sublinear (i.e., convex and homogeneous) function. The set

$$\partial p = \{x' \in (\mathbb{R}^n)' \mid x' \cdot x \leq p(x) \ \forall x \in \mathbb{R}^n\}$$

is the *subdifferential* of p. One can easily check that ∂p is a convex set, and it is closed because it is the intersection of the closed half-spaces $\{x' \in (\mathbb{R}^n)' \mid x' \cdot x \leq p(x)\}$, $x \in \mathbb{R}^n$. Thereby the operator ∂ is defined, which associates with a sublinear function on $\mathbb{R}^n$ a convex closed set in $(\mathbb{R}^n)'$. In particular, with a linear function $l\colon x \to x' \cdot x$ it associates the set $\partial l = \{x'\}$ consisting of a single element.

Note that if p is a sublinear proper and closed function, then $\partial p \neq \emptyset$. Indeed, according to the aforesaid, p^* is also a proper function; hence there exists a point $x' \in (\mathbb{R}^n)'$ where p is finite. In view of homogeneity, $p(x') = 0$. This means that $x' \cdot x \leq p(x)$ for all x, i.e., $x' \in \partial p$.

EXERCISE 3. Determine the subdifferentials of the functions

$$f(x_1, \ldots, x_n) = \Big(\sum_{i=1}^{n} x_i^2\Big)^{1/2},$$
$$f(x_1, \ldots, x_n) = \max_{1 \leq i \leq n} |x_i|,$$
$$f(x_1, \ldots, x_n) = \max_{1 \leq i \leq n} x_i.$$

In the next section we will give the definition of the subdifferential for any convex function at a fixed point (according to which the above definition provides the subdifferential of p at zero), which is a natural extension of the notion of derivative at a given point. A convex function f is approximated locally (e.g., in a neighborhood of a given point x_0) by a sublinear function p (like a smooth function approximated by a linear one), and the subdifferential of f at x_0 is equal to that of p. Hence it is natural to take the above definition as the initial one.

C) *Support operator.* Let A be a convex subset of $\mathbb{R}^n$. The function $s\colon (\mathbb{R}^n)' \to \overline{\mathbb{R}}$ defined by the relation

$$sA(x') = \sup_{x \in A} x' \cdot x$$

(if $A = \emptyset$, we set $sA(x') \equiv -\infty$) is *the support function* of A. It is easily seen that s is a closed sublinear function; hence this definition determines an operator s from the class of convex sets in $\mathbb{R}^n$ to the set of closed sublinear functions on $(\mathbb{R}^n)'$. In particular, if A consists of a single point, $A = \{x\}$, then $sA\colon x' \to x' \cdot x$ is a linear function.

EXERCISE 4. Determine the support functions of the following sets: $\{x \in \mathbb{R}^n \mid |x| \leq 1\}$, $\{x = (x_1, \ldots, x_n)^T \in \mathbb{R}^n \mid \max_{1 \leq i \leq n} |x_i| \leq 1\}$, $\{x \in \mathbb{R}^n \mid \sum_{1 \leq i \leq n} |x_i| \leq 1\}$.

D) *Polar.* Let A be a nonempty convex subset of $\mathbb{R}^n$. The set

$$A^\circ = \{x' \in (\mathbb{R}^n)' \mid x' \cdot x \leq 1 \ \forall x \in A\}$$

is the *polar* of the set A. Obviously, A° is a closed zero-convex set (i.e., a convex closed set containing the origin). Thereby the operator $^\circ$ which associates with a convex set in $\mathbb{R}^n$ a closed zero-convex set in $(\mathbb{R}^n)'$ is defined. In particular, if A consists of a single point, $A = \{x\}$, then the polar is the half-space $\{x' \in (\mathbb{R}^n)' \mid x' \cdot x \leq 1\}$.

One can apply the operator $^\circ$ to the set A° once again. The resulting set $A^{\circ\circ} = (A^\circ)^\circ \subset \mathbb{R}^n$ is the *bipolar* of A.

EXERCISE 5. Determine the polars of the following sets in the plane: $\{(-1,-1),(-1,1),(1,-1),(1,1)\}$, $\{(x_1,x_2) \in \mathbb{R}^2 \mid x_1^2 + x_2^2 \leq 1\}$.

E) *Conjugation operator for cones.* Let C be a convex cone in $\mathbb{R}^n$. The set

$$C' = \{x' \in (\mathbb{R}^n)' \mid x' \cdot x \geq 0, \ \forall x \in C\}$$

is the *conjugate cone* of C. Clearly, C' is a convex closed cone, and thereby the operator $'$ is defined acting from the class of convex cones in $\mathbb{R}^n$ to the class of convex closed cones in $(\mathbb{R}^n)'$. In particular, if the cone is a ray passing through a point x, then the conjugate cone is the half-space $\{x' \in (\mathbb{R}^n)' \mid x' \cdot x \geq 0\}$.

A repeated application of the operator $'$ results in the set $C'' = (C')' \subset \mathbb{R}^n$ called the *bi-conjugate* cone of C.

F) *Annihilator of a subspace.* Let L be a subspace of $\mathbb{R}^n$. The set

$$L^\perp = \{x' \in (\mathbb{R}^n)' \mid x' \cdot x = 0, \ \forall x \in L\}$$

is the *annihilator* of the subspace L. It is clear that $L^\perp$ is a subspace in $(\mathbb{R}^n)'$, so that $^\perp$ is an operator which associates with a subspace of $\mathbb{R}^n$ a subspace of $(\mathbb{R}^n)'$. In particular, the annihilator of the straight line passing through the point x is the hyperplane $\{x' \in (\mathbb{R}^n)' \mid x' \cdot x = 0\}$.

A repeated application of the operator $^\perp$ yields the subspace $L^{\perp\perp} = (L^\perp)^\perp \subset \mathbb{R}^n$ called the *bi-annihilator* of the subspace L.

Besides the operators listed above we will need two more operators, which have already been discussed. They map the subsets of $\mathbb{R}^n$ (or $(\mathbb{R}^n)'$) to functions on $\mathbb{R}^n$ (or $(\mathbb{R}^n)'$). Let us recall their definitions.

Indicator operator. Let A be a subset of $\mathbb{R}^n$. We associate with A the function δA on $\mathbb{R}^n$ (called the *indicator function* of A) by the following rule: $\delta A(x) \equiv +\infty$ if $A = \emptyset$; otherwise, if $A \neq \emptyset$, then $\delta A(x) = 0$ for $x \in A$ and $\delta A(x) = +\infty$ for $x \notin A$.

It is clear that δA is a convex function whenever A is convex.

Minkowski's operator. Let A be a nonempty subset of $\mathbb{R}^n$. We associate with A the function μA on $\mathbb{R}^n$ (called *Minkowski's function* of the set A) by the formula $\mu A(x) = \inf\{\lambda > 0 \mid x \in \lambda A\}$.

EXERCISE 6. Determine Minkowski's function of the triangle in the plane with vertices at the points $(1,0), (-1/2, \sqrt{3}/2), (-1/2, -\sqrt{3}/2)$.

EXERCISE 7. Find the function conjugate with Minkowski's function.

THEOREM (on convex duality).

(1) *Let* $f\colon \mathbb{R}^n \to \mathbb{R} \cup \{+\infty\}$. *Then the following statements are equivalent*:
  (1.1) f *is convex and closed*;
  (1.2) $f^{**} = f$;
  (1.3) f *is the pointwise supremum of affine functions that do not exceed* f.

(2) *Let* $p\colon \mathbb{R}^n \to \overline{\mathbb{R}}$ *be a proper sublinear function. Then the following statements are equivalent*:
  (2.1) p *is closed*;
  (2.2) $s\partial p = p$;
  (2.3) p *is the pointwise supremum of linear functions that do not exceed* p.

(3) *Let* A *be a nonempty subset of* $\mathbb{R}^n$. *Then the following statements are equivalent*:
  (3.1) A *is convex and closed*;
  (3.2) $\partial s A = A$;
  (3.3) A *is the intersection of all half-spaces containing* A.

(4) *Let* A *be a nonempty subset of* $\mathbb{R}^n$. *Then the following statements are equivalent*:
  (4.1) A *is convex, closed, and contains the origin*;
  (4.2) $A^{\circ\circ} = A$;
  (4.3) A *is the intersection of all half-spaces of the form* $\{x \in \mathbb{R}^n \mid x' \cdot x \le 1,\ x' \in (\mathbb{R}^n)'\}$ *containing* A.

(5) *Let* C *be a nonempty subset of* $\mathbb{R}^n$. *Then the following statements are equivalent*:
  (5.1) C *is a convex closed cone*;
  (5.2) $C'' = C$;
  (5.3) C *is the intersection of all half-spaces of the form* $\{x \in \mathbb{R}^n \mid x' \cdot x \le 0,\ x' \in (\mathbb{R}^n)'\}$ *containing* C.

(6) *Let L be a nonempty subset of $\mathbb{R}^n$. Then the following statements are equivalent:*
 (6.1) *L is a subspace;*
 (6.2) *$L^{\perp\perp} = L$;*
 (6.3) *L is the intersection of all hyperplanes of the form $\{x \in \mathbb{R}^n \mid x' \cdot x = 0,\ x' \in (\mathbb{R}^n)'\}$ containing L.*

(7) *Let M be a nonempty subset of $\mathbb{R}^n$. Then the following statements are equivalent:*
 (7.1) *M is an affine subspace;*
 (7.2) *M is the intersection of all affine subspaces of the form $\{x \in \mathbb{R}^n \mid x' \cdot x = x' \cdot x_0,\ x' \in L^{\perp},\ x_0 \in M\}$ containing M, where L is the subspace from which M is obtained by translation.*[1]

PROOF. All the assertions are proved in a unified way. If D is some convex object (the epigraph of a convex function or a convex set) and d is one of the duality operators (*, $^\circ$, $'$, $^\perp$), then it is seen from the definitions that D^{dd} is a convex closed set and $D \subset D^{dd}$. If D is not closed and $D \neq D^{dd}$, then there is an element in $D^{dd} \setminus D$ whose separation from D (by the second separability theorem) immediately leads to a contradiction; hence $D^{dd} = D$. The assertions $s\partial p = p$ and $\partial sA = A$ are proved similarly. Nevertheless, in order not to repeat the same arguments, we deduce some assertions from those proved previously.

Assertion (1) is usually referred to as the *Fenchel–Moreau theorem*.[2]

(1.1) $\Longrightarrow$ (1.3). If $f \equiv +\infty$, then the implication is obvious. Assuming otherwise, we first outline the subsequent proof. If $f(x_0) < +\infty$ for some $x_0 \in \mathbb{R}^n$, then for any $\varepsilon > 0$ the point $(x_0, f(x_0) - \varepsilon)$ does not belong to epi f; hence it can be strictly separated from epi f. But the separating hyperplane is the graph of an affine function which does not exceed f. Since ε is arbitrary, this implies that $f(x_0)$ is the supremum at x_0 of affine functions not exceeding f. If $f(\overline{x}_0) = +\infty$, then for any number $\overline{\alpha}_0$ we can construct an affine function not

---

[1] In fact, for any affine subspace M which does not contain the origin, one can define the duality operator $\cdot$ by the equality $M^{\cdot} = \{x' \in (\mathbb{R}^n)' \mid x' \cdot x = 1,\ \forall x \in M\}$; its extension to subspaces passing through the origin leads to duality in projective geometry, but we will not tackle this matter here.

[2] A geometric proof of the Fenchel–Moreau theorem for functions of one variable was given in the Introduction.

exceeding f with value at $\overline{x}_0$ greater than this number, which will prove the assertion in this case. Now we proceed to precise arguments.

Let $x_0 \in \mathbb{R}^n$ and $\alpha_0 < f(x_0)$. Then $(x_0, \alpha_0) \notin \operatorname{epi} f$. By assumption, the set $\operatorname{epi} f$ is convex and closed; hence by the second separability theorem there exist $x' \in (\mathbb{R}^n)'$ and $\gamma \in \mathbb{R}$ such that

$$x' \cdot x_0 + \alpha_0 \gamma > \sup_{(x,\alpha)\in \operatorname{epi} f} (x' \cdot x + \alpha\gamma). \qquad (i)$$

Now it is easily seen that $\gamma \leq 0$.

Let $f(x_0) < +\infty$. In this case $\gamma < 0$. Indeed, putting the point $(x_0, f(x_0))$ into the right-hand side of (i), we obtain by elementary calculations that $\gamma(\alpha_0 - f(x_0)) > 0$. But $\alpha_0 - f(x_0) < 0$; hence $\gamma < 0$. We may assume that $\gamma = -1$ (dividing inequality (i) by $-\gamma$ if necessary). Then, denoting the right-hand side of (i) by c, we can rewrite this inequality as

$$x' \cdot x_0 - c > \alpha_0, \quad x' \cdot x - c \leq \alpha, \quad \forall (x, \alpha) \in \operatorname{epi} f. \qquad (ii)$$

For the affine function $p(x) = x' \cdot x - c$ we have $p(x) \leq f(x)$ for all $x \in \mathbb{R}^n$ (if $f(x) = +\infty$, this is obvious, and if $f(x) < +\infty$, this follows from the second inequality in (ii) with $\alpha = f(x)$). Moreover, (ii) implies that $\alpha_0 < p(x_0) \leq f(x_0)$, and taking α_0 arbitrarily close to $f(x_0)$ we obtain that f on $\operatorname{dom} f$ is the supremum of affine functions not exceeding f.

Now let $f(\overline{x}_0) = +\infty$. We have to verify that for any $\overline{\alpha}_0 \in \mathbb{R}$ there is an affine function p such that $p(x) \leq f(x)$ for any x and $p(\overline{x}_0) > \overline{\alpha}_0$. If $\gamma < 0$ in (i) (as before, we assume then that $\gamma = -1$), we see from (ii) that $p(\overline{x}_0) > \overline{\alpha}_0$. In the case $\gamma = 0$ we can rewrite (i) as

$$x' \cdot \overline{x}_0 - c > 0, \quad x' \cdot x - c \leq 0, \quad \forall (x, \alpha) \in \operatorname{epi} f. \qquad (ii')$$

According to the aforesaid there exists an affine function p_0 not exceeding f. For an arbitrary $\mu > 0$ consider the affine function $p_\mu(x) = p_0(x) + \mu(x' \cdot x - c)$. The second inequality in (ii') implies that p_μ is also no greater than f on $\operatorname{dom} f$ for any $\mu > 0$, while the first inequality implies that $p_\mu(\overline{x}_0) = p(\overline{x}_0) + \mu(x' \cdot \overline{x}_0 - c) > \overline{\alpha}_0$ for sufficiently large μ.

(1.3) $\Longrightarrow$ (1.2). Let $x' \in (\mathbb{R}^n)'$ and $\alpha \in \mathbb{R}$ be such that $x' \cdot x - \alpha \leq f(x)$ for all $x \in \mathbb{R}^n$. This is equivalent to $\alpha \geq \sup_{x\in\mathbb{R}^n}(x' \cdot x - f(x)) =$

$f^*(x')$. Since f is the supremum of affine functions, this implies that

$$f(x) = \sup_{\substack{x' \in (\mathbb{R}^n)', \alpha \in \mathbb{R} \\ \alpha \geq f^*(x')}} (x' \cdot x - \alpha). \tag{iii}$$

Since f is not equal to $+\infty$ identically, f^* is nowhere equal to $-\infty$; hence we can put $f^*(x')$ for α in (iii). Then

$$f(x) = \sup_{x' \in (\mathbb{R}^n)'} (x' \cdot x - f^*(x')) = f^{**}(x).$$

(1.2) $\Longrightarrow$ (1.1). This implication follows directly from the definition of the conjugate function.

(2) In this part and part (3) of the proof we use two statements which follow directly from the definitions:

(a) if p is a sublinear function on $\mathbb{R}^n$, then $p^* = \delta\partial p$;
(b) if A is a convex subset of $\mathbb{R}^n$, then $(\delta A)^* = sA$.

(2.1) $\Longrightarrow$ (2.2). $p \overset{(1.2)}{=} p^{**} \overset{(a)}{=} (\delta\partial p)^* \overset{(b)}{=} s\partial p$.
(2.2) $\Longrightarrow$ (2.3). The equality (2.2) implies that

$$p(x) = \sup_{x' \in \partial p} x' \cdot x$$

and $x' \cdot x \leq p(x)$ for any $x \in \mathbb{R}^n$, which is equivalent to (2.3).

(2.3) $\Longrightarrow$ (2.1). The epigraph of a support function is the intersection of closed half-spaces; hence p is a closed function.

(3) (3.1) $\Longrightarrow$ (3.2). $\delta A \overset{(1.2)}{=} (\delta A)^{**} \overset{(b)}{=} (sA)^* \overset{(a)}{=} \delta\partial sA$.
(3.2) $\Longrightarrow$ (3.3). The equality (3.2) implies that

$$A = \bigcap_{x' \in (\mathbb{R}^n)'} \{\, x \in \mathbb{R}^n \mid x' \cdot x \leq sA(x') \,\}. \tag{1}$$

Let $x' \in (\mathbb{R}^n)'$. If for some $\gamma \in \mathbb{R}$ the half-space $H_+(x', \gamma) = \{x \in \mathbb{R}^n \mid x' \cdot x \leq \gamma\}$ contains A, then A contains the half-space $\{x \in \mathbb{R}^n \mid x' \cdot x \leq sA(x')\} \subset H_+(x', \gamma)$. If $H_+(x', \gamma)$ contains A for no γ, then $sA(x') = +\infty$; hence $\{x \in \mathbb{R}^n \mid x' \cdot x \leq sA(x')\} = \mathbb{R}^n$. Thus the intersection of all half-spaces containing A coincides with the intersection in the right-hand side of (ii) and hence with A.

(3.3) $\Longrightarrow$ (3.1). A is convex and closed as an intersection of convex closed sets.

(4) (4.1) $\Longrightarrow$ (4.2). Clearly, $A \subset A^{\circ\circ}$. Let us prove the opposite inclusion. Suppose $x_0 \in A^{\circ\circ}$ and $x_0 \notin A$. Then by the second separability theorem there exists a vector $x' \in (\mathbb{R}^n)$, $x' \neq 0$, such

that $\sup_{x\in A} x'\cdot x < x'\cdot x_0$. Since $0 \in A$, we can assume that the left-hand side of the latter inequality is positive and (multiplying x' by a positive constant) set it to be one. Then $x'\cdot x \leq 1$ for any $x \in A$, i.e., $x' \in A^\circ$. But $x'\cdot x_0 > 1$; hence $x_0 \notin A^{\circ\circ}$, which contradicts the assumption.

(4.2) $\Longrightarrow$ (4.3). The equality (4.2) implies that

$$A = \bigcap_{x'\in A^\circ} \{\, x \in \mathbb{R}^n \mid x'\cdot x \leq 1 \,\}. \tag{2}$$

The subsequent reasoning is the same as in the previous case.

(5) (5.1) $\Longrightarrow$ (5.2). Clearly $0 \in C$. Since the cone C along with any x contains the element tx for any $t > 0$, it is easily seen that $C^\circ = -C'$. Next, it is obvious that $(-A)^\circ = -A^\circ$; hence $C'' = -(-C^\circ)^\circ = C^{\circ\circ} \overset{(4.2)}{=} C$.

(5.2) $\Longrightarrow$ (5.3). The equality (5.2) implies that

$$C = \bigcap_{x'\in C'} \{\, x \in \mathbb{R}^n \mid x'\cdot x \leq 0 \,\}. \tag{3}$$

The subsequent reasoning is the same as in the previous case.

(6) (6.1) $\Longrightarrow$ (6.2). We have already pointed out that subspaces in $\mathbb{R}^n$ are closed sets. Clearly each subspace is a cone. Since the subspace L with any x contains tx for any $t \in \mathbb{R}$, we have $L' = L^\perp$. Thus $L^{\perp\perp} = L'' \overset{(5.2)}{=} L$.

(6.2) $\Longrightarrow$ (6.3). The equality (6.2) implies that

$$L = \bigcap_{x'\in L^\perp} \{\, x \in \mathbb{R}^n \mid x'\cdot x = 0 \,\}. \tag{4}$$

The subsequent reasoning is the same as in the previous case.

(7) (7.1) $\Longrightarrow$ (7.2). Since any affine subspace is a convex closed set, we can use (2.2). Let $M = x_0 + L$, where $x_0 \in M$ and L is a subspace. One can easily check that $sM = x'\cdot x_0 + \delta L^\perp$, so that $\partial sM = \{x \in \mathbb{R}^n \mid x'\cdot x \leq x'\cdot x_0,\ \forall x' \in L^\perp\}$. By (2.2) $M = \partial sM$; hence the inequalities $x'\cdot x \leq x'\cdot x_0$ are in fact the equalities. Indeed, the inequality $x'\cdot \overline{x} < x'\cdot x_0$ for some $\overline{x} \in M$ would contradict the fact that $x' \in L^\perp$, since $\overline{x} - x_0 \in L$. Hence we can write

$$M = \bigcap_{x'\in L^\perp,\ x_0\in M} \{\, x \in \mathbb{R}^n \mid x'\cdot x = x'\cdot x_0 \,\} \tag{5}$$

which proves (7.2). □

Note that relations (1) through (5) mean correspondingly that convex closed sets, and only them, are solutions to systems of nonhomogeneous inequalities; closed zero-convex sets, and only them, are solutions to systems of nonhomogeneous inequalities of the form $a \cdot x \leq 1$; convex closed cones, and only them, are solutions to systems of homogeneous inequalities; affine manifolds, and only them, are solutions to systems of nonhomogeneous equations.

3. Convex calculus

- In this section we expose convex calculus, which is a collection of formulas for the action of duality operators (conjugation for functions, subdifferential, support, and polar for sets, conjugation for cones, and annihilator for subspaces) on elements that are themselves obtained from a pair of elements by some operations on sets or functions (e.g., the sum or intersection of convex sets, or the maximum or sum of convex functions), as well as on elements that are images or inverse images of sets or functions under a linear mapping.

The major part of this section is devoted to subdifferential calculus, which is the most essential part of convex calculus. Subdifferential calculus in convex analysis is an analog of differential calculus, where the role of derivative is played by the subdifferential.

DEFINITION 3. Let $f: \mathbb{R}^n \to \overline{\mathbb{R}}$ with $f(x_0)$ finite at $x_0 \in \mathbb{R}^n$. The set (possibly empty)

$$\partial f(x_0) = \{x' \in (\mathbb{R}^n)' \mid f(x) - f(x_0) \geq x' \cdot (x - x_0),\ \forall x \in \mathbb{R}^n\}$$

is the *subdifferential* of f at the point x_0.

The elements of $\partial f(x_0)$ are called *subgradients* of f at x_0. It is clear that $x' \in (\mathbb{R}^n)'$ is a subgradient of f at x_0 if and only if there exists $\alpha \in \mathbb{R}$ such that the affine function $x \to x' \cdot x + \alpha$ does not exceed f everywhere and equals $f(x_0)$ at the point x_0.

It is not hard to check that if $f: \mathbb{R}^n \to \overline{\mathbb{R}}$ is convex and differentiable at $x_0 \in \mathbb{R}^n$, then $\partial f(x_0) = \{f'(x_0)\}$. Indeed, it has been shown that the function $t^{-1}((f(x_0) + tx) - f(x_0))$ is monotone decreasing in t for any x; hence $f(x_0+x) - f(x) \geq f'(x_0; x) = f'(x_0) \cdot x$, i.e., $f'(x_0) \in \partial f(x_0)$. Conversely, if $x' \in \partial f(x_0)$, then, by definition,

$f(x_0+tx)-f(x_0)\geq t\,x'\cdot x$ for all $x\in X$ and $t>0$. Dividing this inequality by t and taking the limit as $t\downarrow 0$, we obtain $f'(x_0)\cdot x\geq x'\cdot x$, i.e., $x'=f'(x_0)$. □

Now we state a simple fact playing, however, a very important role in the theory of convex extremal problems. This fact is an analog of the Fermat theorem for convex functions. Let $f:\mathbb{R}^n\to\overline{\mathbb{R}}$ be a convex function. Consider the extremal problem

$$f(x)\to\min,\quad x\in\mathbb{R}^n.$$

This is a minimization problem of a convex function without constraints.

THEOREM (analog of Fermat's theorem). *A point $\widehat{x}\in\mathbb{R}^n$ affords a minimum in the problem without constraints if and only if $0\in\partial f(\widehat{x})$.*

PROOF. The function f attains the minimum at the point $\widehat{x}\in\mathbb{R}^n$ (obviously, $f(\widehat{x})<+\infty$) if and only if $f(x)-f(\widehat{x})\geq 0=0\cdot(x-\widehat{x})$ for all $x\in\mathbb{R}^n$, which is equivalent to the inclusion $0\in\partial f(\widehat{x})$. □

Now we illustrate the calculation of subdifferentials by some examples.

EXAMPLE 1. $f(x)=e^x$, $x\in\mathbb{R}$. This is a convex smooth function. By what we proved above, $\partial f(x)=e^x$.

Here is an example of a convex function (again on $\mathbb{R}$) with empty subdifferential at a point. Let $f(x)=x\log x$ for $x>0$, $f(0)=0$, and $f(x)=+\infty$ for $x<0$. Since there is no $k\in\mathbb{R}$ to satisfy the inequality $x\log x\geq kx$ for all $x>0$, we see that $\partial f(0)=\emptyset$.

EXAMPLE 2. $f(x)=|x|$, $x\in\mathbb{R}$. It is again a convex function. It is clear that $k\in\partial f(0)$ if and only if $kx\leq|x|$ for all $x\in\mathbb{R}$, i.e., $\partial f(0)=[-1,1]$. It $x_0\neq 0$, then, obviously, $k\in\partial f(x_0)$ if and only if $k=\operatorname{sign}x_0$, i.e., $k\in\partial f(0)$ and $kx_0=|x_0|$.

In the next example this situation is extended to arbitrary sublinear functions.

EXAMPLE 3. Let $p\colon\mathbb{R}^n\to\mathbb{R}$ be a sublinear function. We will denote its subdifferential at zero simply by ∂p, i.e., $\partial p=\{x'\in(\mathbb{R}^n)'\mid x'\cdot x\leq p(x),\ \forall x\in\mathbb{R}^n\}$. If p is continuous, then, for any $x_0\in\mathbb{R}^n$,

$$\partial p(x_0)=\{x'\in(\mathbb{R}^n)'\mid x'\in\partial p,\ x'\cdot x_0=p(x_0)\}.$$

Indeed, let $x' \in \partial p(x_0)$. Then $p(x) - p(x_0) \geq x' \cdot (x - x_0)$ for any $x \in \mathbb{R}^n$. In this inequality we let $x = 0$ and then $x = 2x_0$ to obtain $p(x_0) = x' \cdot x_0$. Next, if $\lambda > 0$, then $p(\lambda x + x_0) - p(x_0) \geq x' \cdot \lambda x$, or $p(x + \lambda^{-1}x_0) - \lambda^{-1}p(x_0) \geq x' \cdot x$. Letting $\lambda \to \infty$ we obtain that $p(x) \geq x' \cdot x$, i.e., $x' \in \partial p$. Conversely, let $x' \in \partial p$ and $p(x_0) = x' \cdot x_0$. Then $p(x) - p(x_0) \geq x' \cdot (x - x_0)$ for any $x \in \mathbb{R}^n$; hence $x' \in \partial p(x_0)$.□

EXAMPLE 4. Let δA be the indicator function of a set $A \in \mathbb{R}^n$. Then it follows directly from the definitions that for any $x_0 \in A$

$$\begin{aligned}\partial\delta A(x_0) &= \{x' \in (\mathbb{R}^n)' \mid x' \cdot (x - x_0) \leq 0, \ \forall x \in A\} \\ &= \{x' \in (\mathbb{R}^n)' \mid x' \cdot x_0 = sA(x')\}.\end{aligned}$$

Clearly, this is a cone in $(\mathbb{R}^n)'$. If $A = L$ is a subspace in $\mathbb{R}^n$, then $\partial\delta L(x_0) = \{x' \in (\mathbb{R}^n)' \mid x' \cdot x = 0, \ \forall x \in L\} = L^\perp$ is the annihilator of L.

Now we will state the main properties of a subdifferential.

THEOREM (on existence and structure of a subdifferential). *Let $f \colon \mathbb{R}^n \to \overline{\mathbb{R}}$ be a convex function finite at $x_0 \in \mathbb{R}^n$. Then*

1) *$\partial f(x_0)$ is a convex closed set;*
2) *$\partial f(x_0) = \partial f'(x_0; \cdot)$;*
3) *if f is continuous at x_0, then $\partial f(x_0)$ is a nonempty convex compact set.*

PROOF. 1) Since $\partial f(x_0)$ is the intersection of closed half-spaces $\{x' \in (\mathbb{R}^n)' \mid x' \cdot (x - x_0) \leq f(x) - f(x_0)\}$ over all $x \in \mathbb{R}^n$, this is a convex closed set (possibly empty).

2) Let $x' \in \partial f(x_0)$. Then for any $x \in \mathbb{R}^n$ and $t > 0$ we have $x' \cdot x = t^{-1}x' \cdot tx \leq t^{-1}(f(x_0 + tx) - f(x_0))$. Taking the limit as $t \to 0$ we obtain that $x' \in \partial f'(x_0; 0)$. Conversely, if $x' \in \partial f'(x_0; 0)$, then we have from the definitions $x' \cdot (x - x_0) \leq f'(x_0; x - x_0) \leq f(x_0 + x - x_0) - f(x_0) = f(x) - f(x_0)$ for all $x \in \mathbb{R}^n$, i.e., $x' \in \partial f(x_0)$.

3) First we note that continuity of f at x_0 implies that $f'(x_0; \cdot)$ is a closed proper function. Indeed, the function $x \to f(x + x_0) - f(x_0)$ is bounded in some neighborhood of the origin, and since $f'(x_0; x) \leq f(x + x_0) - f(x_0)$, the function $f'(x_0; \cdot)$ is also bounded in this neighborhood. Therefore it is continuous at the origin and hence everywhere (hence also bounded everywhere). Continuity of a function implies its closedness.

Now we will show that $\partial f(x_0) \neq \emptyset$. It suffices to verify that $\partial p \neq \emptyset$. Denote $p = f'(x_0, \cdot)$. It $\partial p = \emptyset$, then for any $x' \in (\mathbb{R}^n)'$

there is $x \in \mathbb{R}^n$ such that $x' \cdot x - p(x) > 0$; hence $p^* \equiv +\infty$. But then $p^{**} \equiv -\infty$, which contradicts the Fenchel–Moreau theorem.

In view of continuity of f at x_0 there is $\delta > 0$ such that $f(x_0 + x) - f(x) \leq 1$ for all $x \in B_\delta(0)$. If $x' \in \partial f(x_0)$, then such x fulfills the inequality $x' \cdot x \leq f(x_0 + x) - f(x) \leq 1$. This implies that $\partial f(x_0)$ is contained in the ball of radius $\sqrt{n}/\delta$. Hence $\partial f(x_0)$ is a nonempty compact set. □

Now we will prove two main theorems of convex calculus (see Ioffe and Tikhomirov (1979)).

THEOREM (Moreau–Rockafellar). *Let $f_i\colon \mathbb{R}^n \to \overline{\mathbb{R}}$, $i = 1, 2$, be two convex proper functions. Suppose there exists a point where both functions are finite and at least one of them is continuous. Then*

$$\partial(f_1 + f_2)(x) = \partial f_1(x) + \partial f_2(x)$$

for all $x \in \operatorname{dom} f_1 \cap \operatorname{dom} f_2$.

PROOF. The previous theorem and the obvious fact that $(f_1 + f_2)'(x;\cdot) = f_1'(x;\cdot) + f_2'(x;\cdot)$, $x \in \mathbb{R}^n$, imply that it suffices to prove the theorem for sublinear functions $p_i = f_i'(x;\cdot)$, $i = 1, 2$. We will restrict ourselves to the case where one of them, say p_1, is closed and the other one is continuous (hence closed as well). In this case ∂p_1 is a nonempty convex closed set (which follows from the proof of the previous theorem), and ∂p_2 is a nonempty convex compact set; hence the set $\partial p_1 + \partial p_2$ is closed (indeed, if $x_k^1 + x_k^2 \to x \in \operatorname{cl}(\partial p_1 + \partial p_2)$, $x_k^i \in \partial p_i$, $i = 1, 2$, then by compactness of ∂p_2 one can select from $\{x_k^2\}$ a subsequence $\{x_{k_l}^2\}$ such that $x_{k_l} \to x^2 \in \partial p_2$. But then $x_{k_l}^1 \to x^1 = x - x^2$. Since ∂p_1 is closed, $x^1 \in \partial p_1$; hence $x = x^1 + x^2 \in \partial p_1 + \partial p_2$). We will also need the easily verifiable relation

$$s(A_1 + A_2) = sA_1 + sA_2, \tag{$*$}$$

which is valid for arbitrary sets A_1 and A_2 in $\mathbb{R}^n$.

Now applying successively the second statement of the theorem on convex duality (see Section 2), $(*)$, and the third statement of that theorem, we obtain

$$\partial(p_1 + p_2) = \partial(s\partial p_1 + s\partial p_2) = \partial s(\partial p_1 + \partial p_2) = \partial p_1 + \partial p_2. \qquad \square$$

THEOREM (Dubovitskii–Milyutin). *Let $f_i\colon \mathbb{R}^n \to \overline{\mathbb{R}}$, $i = 1, 2$, be convex functions continuous at a point $x \in \mathbb{R}^n$ and let $f_1(x) = f_2(x)$. Then*

$$\partial \max(f_1, f_2)(x) = \operatorname{co}(\partial f_1(x) \cup \partial f_2(x)).$$

PROOF. Since $(\max(f_1, f_2))'(x; \cdot) = \max(f_1'(x; \cdot), f_2'(x; \cdot))$ (which can be verified trivially) it again suffices to prove the theorem for sublinear functions $p_i = f_i'(x; \cdot)$, $i = 1, 2$. Since f_i, $i = 1, 2$, are continuous at x, by the theorem on existence and structure of the subdifferential the sets ∂p_i, $i = 1, 2$, are compact; hence the set $\operatorname{co}(\partial p_1 \cup \partial p_2)$ is compact as well (indeed, if $(1 - \alpha_k)x_k^1 + \alpha_k x^2 \to x \in \operatorname{cl} \operatorname{co}(\partial p_1 \cup \partial p_2)$, $x_k^i \in \partial p_i$, $i = 1, 2$, $0 \le \alpha_k \le 1$, then, as in the previous theorem, one can select a subsequence $\{x_{k_l}\}$ such that $(1 - \alpha_k)x_k^1 + \alpha_k x^2 \to (1 - \alpha)x^1 + \alpha x^2$, where $x^i \in \partial p_i$, $i = 1, 2$, and $0 \le \alpha \le 1$. Thus $x \in \operatorname{co}(\partial p_1 \cup \partial p_2)$). We will also need the following easily verifiable equality

$$s(\operatorname{co}(A_1 \cap A_2)) = \max(sA_1, sA_2), \tag{**}$$

which is valid for any $A_i \subset \mathbb{R}^n$, $i = 1, 2$. Now applying successively the second statement of the theorem on convex duality, (**), and the third statement of that theorem, we obtain

$$\begin{aligned}\partial \max(p_1, p_2) &= \partial \max(s\partial p_1, s\partial p_2)\\ &= \partial s(\operatorname{co}(\partial p_1 \cup \partial p_2)) = \operatorname{co}(\partial p_1 \cup \partial p_2). \qquad \square\end{aligned}$$

EXERCISE 1. Find the subdifferentials of the following functions: $\max(e^{-x}, e^x)$ at $x = 0$, $\max(-\log x, x - 1)$ at $x = 1$, and $\max(x, x^2)$ at $x = 0$ and $x = 1$.

It is clear that the last two theorems can be extended by induction to any finite number of functions. The following result extends the Dubovitskii–Milyutin theorem.

THEOREM (Levin (1985); subdifferential form of the clean-up theorem).[3] *Let T be a compact set in $\mathbb{R}^m$. Suppose a function $f\colon T \times \mathbb{R}^n \to \mathbb{R}$ is such that the mapping $f(t, \cdot)\colon \mathbb{R}^n \to \mathbb{R}$ is convex for each $t \in T$ and the mapping $f(\cdot, x)\colon T \to \mathbb{R}$ is continuous for each $x \in \mathbb{R}^n$. Let $x' \in \partial \max_{t \in T} f(t, \widehat{x})$ for some $\widehat{x} \in \mathbb{R}^n$. Then*

[3]The very idea of "clean-up" appeared in the works of de la Vallée-Poussin on approximation theory. It was put in a more complete form by L. G. Shnirel'man.

there is a positive integer $r \le n+1$ and points $t_i \in T$, $i = 1, \dots, r$, such that

$$x' \in \operatorname{co} \bigcup_{i=1}^{r} \partial_x f(t_i, x_0), \qquad \max_{1 \le i \le r} f(t_i, \widehat{x}) = f(t_1, \widehat{x}) = \cdots = f(t_r, \widehat{x}),$$

where $\partial_x f(t, \widehat{x})$ denotes the subdifferential of the function $x \to f(t,x)$ at $\widehat{x}$.

PROOF. The condition $x' \in \partial \max_{t \in T} f(t, \widehat{x})$ is equivalent to $0 \in \partial g(\widehat{x})$, where $g(x) = \max_{t \in T} f(t,x) - x' \cdot x = \max_{t \in T}(f(t,x) - x' \cdot x)$. By Fermat's theorem $\widehat{x}$ is a minimum point of $g(\cdot)$. Then by the clean-up theorem (see Section 11) applied to this function there are a positive integer $r \le n+1$ and points $t_i \in T$, $i = 1, \dots, r$, such that $\widehat{x}$ is a minimum point of the function $\max_{1 \le i \le r} f(t_i, x) - x' \cdot x$. This implies, again by Fermat's theorem, that the subdifferential of this function at $\widehat{x}$ contains zero, or equivalently, $x' \in \partial \max_{1 \le i \le r} f(t_i, \widehat{x})$. Reducing r if necessary, we may assume that the values at $\widehat{x}$ of the functions $f(t_i, \cdot)$, $i = 1, \dots, r$, are equal. Now the required assertion follows directly from the Dubovitskii–Milyutin theorem. □

Formulas of convex calculus. First we list the operations on convex sets and convex functions. We begin with sets. The following operations associate with each pair of convex subsets of $\mathbb{R}^n$ a convex set in the same space:

- convex hull of a union ($\operatorname{co}\cup$): $(A_1, A_2) \to A_1 \operatorname{co}\cup A_2$;
- intersection ($\cap$): $(A_1, A_2) \to A_1 \cap A_2$;
- sum ($+$): $(A_1, A_2) \to A_1 + A_2$;
- Kelley's sum ($\otimes$): $(A_1, A_2) \to A_1 \otimes A_2 = \bigcup_{0 \le \alpha \le 1}((1-\alpha)A_1 \cap \alpha A_2)$.

Let $\Lambda\colon \mathbb{R}^n \to \mathbb{R}^m$ be a linear operator. Then one can associate with each convex set $A \subset \mathbb{R}^n$ ($B \subset \mathbb{R}^m$) a convex set in $\mathbb{R}^m$ ($\mathbb{R}^n$) by the following rule:

- the image of A under the mapping Λ: $\Lambda A = \{y \in \mathbb{R}^m \mid \exists x \in A\colon\ y = \Lambda x\}$;
- the inverse image of B under the mapping Λ: $B\Lambda = \{x \in \mathbb{R}^n \mid \Lambda x \in B\}$.

The following operations on convex functions associate with each pair of convex functions on $\mathbb{R}^n$ a convex function on the same space (for numbers a and b we will usually write $a \vee b$ instead of $\max\{a, b\}$):

- convex hull of minimum ($\operatorname{co}\wedge$): $(f_1, f_2) \to f_1 \operatorname{co}\wedge f_2$ $((f_1 \operatorname{co}\wedge f_2)(x) = \inf\{(1-\alpha)f_1(x_1) + \alpha f_2(x_2) \mid x = (1-\alpha)x_1 + \alpha x_2, 0 \le \alpha \le 1\})$;
- maximum ($\vee$): $(f_1, f_2) \to f_1 \vee f_2$ $((f_1 \vee f_2)(x) = \max(f_1(x), f_2(x)))$;
- convolution ($\oplus$): $(f_1, f_2) \to f_1 \oplus f_2$ $((f_1 \oplus f_2)(x) = \inf\{f_1(x_1) + f_2(x_2) \mid x = x_1 + x_2\})$;
- sum ($+$): $(f_1, f_2) \to f_1 + f_2$ $((f_1 + f_2)(x) = f_1(x) + f_2(x))$;
- Kelley's sum ($\otimes$): $(f_1, f_2) \to f_1 \otimes f_2$ $(f_1 \otimes f_2)(x) = \inf\{f_1(x_1) \vee f_2(x_2) \mid x = x_1 + x_2\}$.

EXERCISE 2. Let $f_1(x) = e^x$, $f_2(x) = e^{-x}$, and $f_3(x) = x^2$. Show that $(f_1 \oplus f_2)(x) = 0$ and $(f_3 \oplus f_2)(x) = x^2/2$.

Moreover, we will introduce two more operations which associate a convex function to a convex function and a linear operator. Let $\Lambda\colon \mathbb{R}^n \to \mathbb{R}^m$ be a linear operator, f a convex function on $\mathbb{R}^m$, and g a convex function on $\mathbb{R}^n$:

- the inverse image of f under the mapping Λ: $(f, \Lambda) \to f\Lambda$ $(f\Lambda(x) = f(\Lambda x))$;
- the image of g under the mapping Λ: $(\Lambda, g) \to \Lambda g$ $(\Lambda g(y) = \inf\{g(x) \mid \Lambda x = y\})$.

The operations on sets thus introduced (except for Kelley's sum) look quite natural. Kelley's sum, as will be seen in the sequel, arises from the reasons of duality. The operations on functions (except for maximum, sum, and inverse image under a linear mapping) look rather artificial, but in fact all these operations result in a natural way from the set operations applied to the epigraphs of the functions. Specifically, if C is a convex subset of $\mathbb{R}^n \times \mathbb{R}$, then it is not hard to check that the function $f\colon \mathbb{R}^n \to \overline{\mathbb{R}}$ defined as $f(x) = \inf\{t \in \mathbb{R} \mid (x,t) \in C\}$ is convex. Now if f_1 and f_2 are convex functions, then one can easily check that the set $C = \operatorname{co}(\operatorname{epi} f_1 \cup \operatorname{epi} f_2)$ corresponds to the convex hull of the minimum and, obviously, the set $C = \operatorname{epi} f_1 \cap \operatorname{epi} f_2$ corresponds to the maximum of these functions. Further, it is not hard to see that $C = \operatorname{epi} f_1 + \operatorname{epi} f_2$ corresponds to the convolution. Using that the epigraphs lie in the product spaces, one can associate with two convex sets $\operatorname{epi} f_1$ and $\operatorname{epi} f_2$ the convex sets $\{(x,t) \in \mathbb{R}^n \times \mathbb{R} \mid (x, t_i) \in \operatorname{epi} f_i,\ i = 1,2\}$ and $\{(x,t) \in \mathbb{R}^n \times \mathbb{R} \mid (x_i, t) \in \operatorname{epi} f_i,\ i = 1,2\}$. One can easily check that the former corresponds to the sum of functions and the latter to Kelley's sum. The image of a function arises from the reasons of duality.

In the following theorem the sign "$=$" means that the equality of the convex objects in question holds without any additional assumptions, while the sign "$\cong$" means that the equality is only valid under certain additional conditions.

THEOREM (the main formulas of convex calculus).

(1) *Let f_1, f_2, g be convex functions on $\mathbb{R}^n$, f a convex function on $\mathbb{R}^m$, $\Lambda\colon \mathbb{R}^n \to \mathbb{R}^m$ a linear operator, and $\Lambda'\colon (\mathbb{R}^m)' \to (\mathbb{R}^n)'$ the operator conjugate with Λ. Then*

1.1 $(f_1 \vee f_2)^* \cong f_1^* \operatorname{co} \wedge f_2^*$; 1.2 $(f_1 \operatorname{co} \wedge f_2)^* = f_1^* \vee f_2^*$;

1.3 $(f_1 \oplus f_2)^* = f_1^* + f_2^*$; 1.4 $(f_1 + f_2)^* \cong f_1^* \oplus f_2^*$;

1.5 $(\Lambda g)^* = g^* \Lambda'$; 1.6 $(f\Lambda)^* \cong \Lambda' f^*$.

(2) *Let A_1, A_2, A be convex subsets of $\mathbb{R}^n$; B a convex subset of $\mathbb{R}^m$; and $\Lambda\colon \mathbb{R}^n \to \mathbb{R}^m$ a linear operator. Then*

2.1 $\delta(A_1 \cap A_2) = \delta(A_1 \otimes A_2) = \delta A_1 + \delta A_2 = \delta A_1 \vee \delta A_2$;

2.2 $\delta(A_1 \operatorname{co}\cup A_2) = \delta A_1 \operatorname{co} \wedge \delta A_2$; 2.3 $\delta(A_1 + A_2) = \delta A_1 \oplus \delta A_2$;

2.4 $\delta(\Lambda A) = \Lambda\, \delta A$; 2.5 $\delta(B\Lambda) = \delta B\, \Lambda$.

(3) *Let A_1, A_2, A, B, Λ be as in the previous item and Λ' the operator conjugate with Λ. Then*

3.1 $s(A_1 \cap A_2) \cong sA_1 \oplus sA_2 \cong s\, A_1 \operatorname{co} \wedge sA_2$;

3.2 $s(A_1 \operatorname{co}\cup A_2) = sA_1 \vee sA_2$; 3.3 $s(A_1 + A_2) = sA_1 + sA_2$;

3.4 $s(A_1 \otimes A_2) \cong sA_1 \otimes sA_2$; 3.5 $s(\Lambda A) = sA\, \Lambda'$;

3.6 $s(B\Lambda) \cong \Lambda'\, sB$.

(4) *Let A_1, A_2, A, B, Λ be as in the previous item. Then*

4.1 $\mu(A_1 \cap A_2) = \mu A_1 \vee \mu A_2$; 4.2 $\mu(A_1 \operatorname{co}\cup A_2) \cong \mu A_1 \oplus \mu A_2$;

4.3 $\mu(A_1 + A_2) \cong \mu A_1 \otimes \mu A_2$; 4.4 $\mu(A_1 \otimes A_2) \cong \mu A_1 + \mu A_2$;

4.5 $\mu(\Lambda A) = \Lambda\, \mu A$; 4.6 $\mu(B\Lambda) = \mu B\, \Lambda$.

(5) *Let A_1, A_2, A, B, Λ be as in the previous item and Λ' the operator conjugate with Λ. Then*

5.1 $(A_1 \cap A_2)^\circ \cong A_1^\circ \operatorname{co}\cup A_2^\circ$; 5.2 $(A_1 \operatorname{co}\cup A_2)^\circ = A_1^\circ \cap A_2^\circ$;

5.3 $(A_1 + A_2)^\circ = A_1^\circ \otimes A_2^\circ$; 5.4 $(A_1 \otimes A_2)^\circ \cong A_1^\circ + A_2^\circ$;

5.5 $(\Lambda A)^\circ = A^\circ \Lambda'$; 5.6 $(B\Lambda)^\circ = \Lambda' B^\circ$.

(6) *Let p_1, p_2 be sublinear functions on $\mathbb{R}^n$, $\Lambda\colon \mathbb{R}^n \to \mathbb{R}^m$ a linear operator, and Λ' the operator conjugate with Λ. Then*

6.1 $\partial(p_1 \vee p_2) \cong \partial p_1 \operatorname{co}\cup \partial p_2$; 6.2 $\partial(p_1 \operatorname{co}\wedge p_2) = \partial p_1 \cap \partial p_2$;
6.3 $\partial(p_1 \oplus p_2) = \partial p_1 \cap \partial p_2$; 6.4 $\partial(p_1 + p_2) \cong \partial p_1 + \partial p_2$;
6.5 $\partial(p\Lambda) \cong \Lambda' \partial(p\Lambda)$.

(7) *Let C and L be a convex cone and a subspace in $\mathbb{R}^n$ respectively. Let $\Lambda\colon \mathbb{R}^n \to \mathbb{R}^m$ be a linear operator and $\Lambda'\colon (\mathbb{R}^m)' \to (\mathbb{R}^n)'$ the operator conjugate with Λ. Then*

7.1 $(C_1 + C_2)' = C_1' \cap C_2'$; 7.2 $(C_1 \cap C_2)' \cong C_1' + C_2'$;
7.3 $(L_1 + L_2)^\perp = L_1^\perp \cap L_2^\perp$; 7.4 $(L_1 \cap L_2)^\perp \cong L_1^\perp + L_2^\perp$;
7.5 $(\Lambda L)^\perp = L^\perp \Lambda'$; 7.6 $(L\Lambda)^\perp \cong \Lambda' L^\perp$;
7.7 $(\Lambda C)' = C'\Lambda'$; 7.8 $(C\Lambda)' \cong \Lambda' C'$.

PROOF. We will not prove all the formulas; the rest is left to the reader. The formulas 1.2, 1.3, 1.5 follow easily from the definitions. Formulas 1.1 and 1.4 are valid under the additional assumption $\operatorname{ri}(\operatorname{dom} f_1) \cap \operatorname{ri}(\operatorname{dom} f_2) \neq \emptyset$, and 1.6 holds provided there is x such that $\Lambda x \in \operatorname{ri}(\operatorname{dom} f)$ (see Rockafellar (1970)).

The formulas in part (2) follow from the definitions, as well as formulas 3.2, 3.3, and 3.5.

Formula 3.1 is valid if $\operatorname{ri} A_1 \cap \operatorname{ri} A_2 \neq \emptyset$. Indeed, using the easily verifiable relation

$$(\delta A)^* = sA \tag{i}$$

(which we encountered already) we have

$$\begin{aligned} s(A_1 \cap A_2) &\overset{(i)}{=} (\delta(A_1 \cap A_2))^* \overset{(2.1)}{=} (\delta A_1 + \delta A_2)^* \\ &\overset{(1.4)}{=} (\delta A_1)^* \oplus (\delta A_2)^* \overset{(i)}{=} sA_1 \oplus sA_2. \end{aligned}$$

Formula 3.6 holds provided there is x such that $\Lambda x \in \operatorname{ri} A$. Indeed, using the same relation we have

$$s(B\Lambda) \overset{(i)}{=} (\delta(B\Lambda))^* \overset{(2.6)}{=} ((\delta B)\Lambda)^* \overset{(1.6)}{=} \Lambda'(\delta B)^* \overset{(i)}{=} \Lambda' sB.$$

Formulas 4.1, 4.2, 4.6 follow easily from the definitions. For the proof of the remaining ones we formulate some auxiliary assertions, which can be established without difficulty: if $\{A_j\}_{j\in J}$ is an arbitrary family of sets, then

$$\mu\Big(\bigcup_{j\in J} A_j\Big) = \inf_{j\in J} \mu A_j; \tag{ii}$$

if A_1 and A_2 are arbitrary sets, then

$$\mu(A_1 \times A_2) = \mu A_1 \vee \mu A_2; \tag{iii}$$

and, finally, if a and b are arbitrary nonnegative numbers, then

$$\inf\{\max(a/\alpha, b/\beta) \mid \alpha > 0, \beta > 0,\ \alpha + \beta = 1\} = a + b. \tag{iv}$$

Let us prove 4.3. We apply formula 4.5 with operator $\Lambda\colon \mathbb{R}^n \times \mathbb{R}^n \to \mathbb{R}^n$, $\Lambda(x_1, x_2) = x_1 + x_2$. Then for any $x \in \mathbb{R}^n$ we have

$$\begin{aligned}
\mu(A_1 + A_2)(x) &= \mu(\Lambda(A_1 \times A_2))(x) \\
&\overset{(4.5)}{=} \inf\{\mu(A_1 \times A_2)(x_1, x_2) \mid x = x_1 + x_2\} \\
&\overset{(iii)}{=} \inf\{(\mu A_1)(x_1) \vee (\mu A_2)(x_2) \mid x = x_1 + x_2\} \\
&= (\mu A_1 \otimes \mu A_2)(x).
\end{aligned}$$

Now we prove 4.4:

$$\begin{aligned}
\mu(A_1 \otimes A_2)(x) &= \mu\Big(\bigcup_{\alpha,\beta>0,\ \alpha+\beta=1} (\alpha A_1 \cap \beta A_2)\Big)(x) \\
&\overset{(ii)}{=} \inf\big\{\mu(\alpha A_1 \cap \beta A_2)(x) \mid \alpha, \beta \geq 0,\ \alpha + \beta = 1\big\} \\
&\overset{(4.1)}{=} \inf\Big\{\max\Big(\frac{\mu A_1(x)}{\alpha}, \frac{\mu A_2(x)}{\beta}\Big) \Big| \alpha, \beta \geq 0,\ \alpha + \beta = 1\Big\} \\
&\overset{(iv)}{=} \mu A_1(x) + \mu A_2(x).
\end{aligned}$$

4. Finite-dimensional convex geometry

- Convex geometry traces its origin to the 19th century in the works of Cauchy, Steiner, and Minkowski. It reached its fullest flower in the 20th century. Nowadays finite-dimensional convexity also attracts considerable interest due to its applications to optimization algorithms and algebraic geometry. In this section we present some classical results in this area.

4.1. Cauchy's theorem on rigidity of convex polyhedra. We begin with the first fundamental result in the history of convex geometry. Cauchy's theorem says that *a convex polyhedron in the three-dimensional space is uniquely, up to translation, determined by its faces.*

Now we formulate this result more precisely. We will say that two polyhedra, M and M', are *equivalent* if there is a correspondence f between their faces, edges, and vertices such that if Γ is a face or an edge of one of them and if Γ_1 is, respectively, an edge or a vertex of Γ, then $f(\Gamma_1)$ is an edge or a vertex of the face or edge $f(\Gamma)$ of the other polyhedron.

THEOREM (Cauchy). *If the respective faces of two equivalent convex polyhedra in the three-dimensional space are isometric, then the polyhedra themselves are isometric.*

Isometric bodies are also called *congruent*, i.e., superposable by translation. Hence Cauchy's theorem can be reformulated as follows: *If the respective faces of two convex polyhedra are congruent, then the polyhedra themselves are congruent.* If we drop the convexity condition, then this assertion fails; see Figure 4.1.

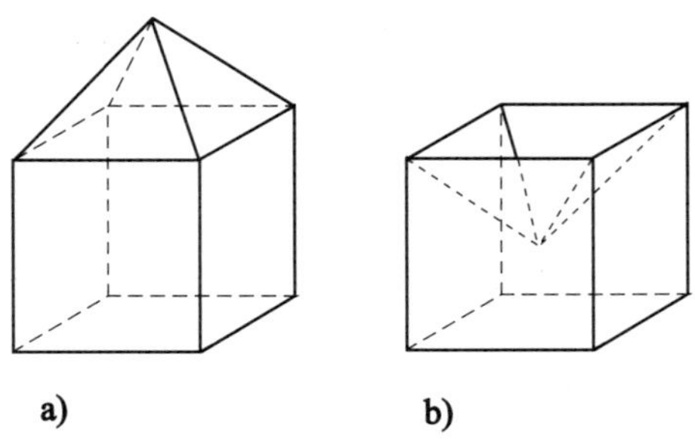

FIGURE 4.1

PROOF. We will deduce Cauchy's theorem from the following geometric lemma.

LEMMA 1. *Suppose the faces of two noncongruent polyhedra can be put in a correspondence so that the planar angles of the corresponding faces are equal. Let the edges of the first polyhedron be marked with + or − if the corresponding bihedral angle at this edge is greater or less than the corresponding bihedral angle of the other polyhedron. Next, we mark each planar angle between two marked edges with* 1 *or* 0 *according to whether the signs on these edges are different or the same. Then there are at least four angles of the first polyhedron marked with* 1.

To explain this idea, imagine a regular tetrahedral angle. Let us press the opposite edges towards each other. Then the corresponding bihedral angles increase, while the other two decrease, so that the edges will be marked with alternate signs, and all of the four planar angles will be marked with 1.

We will deduce Cauchy's theorem from the lemma and then prove the lemma. Assume that not all of the bihedral angles of the two polyhedra are equal. Let us mark the edges of M with the signs $+$ or $-$ if the corresponding bihedral angle of M is greater or less than that of M'. Again, we mark a planar angle between two marked edges with 1 or 0 according to whether the marks of its sides alternate or coincide.

Consider first the case where *all* the edges of the polyhedron are marked.

Let N_0 be the number of vertices, N_1 the number of edges, and N_2 the number of faces of the polyhedra, and let the total number of planar angles of the first polyhedron marked with 1 be N. Denote by a_3 the number of triangular faces, by a_4 that of quadrangular faces, and, in general, by a_k the number of n-gonal faces.

The number of 1-angles in a k-gonal face is even and does not exceed k. Hence

$$N \leq 2a_3 + 4a_4 + 4a_5 + 6a_6 + 6a_7 + \cdots . \qquad (i)$$

Moreover, we have

$$a_3 + a_4 + a_5 + \cdots = N_2 \quad \text{and} \quad 3a_3 + 4a_4 + 5a_5 + \cdots = 2N_1. \qquad (ii)$$

Hence

$$4(N_1 - N_2) \overset{(ii)}{=} 2a_3 + 4a_4 + 6a_5 + \cdots \overset{(i)}{\geq} N. \qquad (iii)$$

By Euler's formula

$$4(N_1 - N_2) = 4N_0 - 8. \qquad (iv)$$

Thus $N \overset{(iii),(iv)}{\leq} 4N_0 - 8 < 4N_0$. On the other hand, by Lemma 1 we have at least four 1-angles adjacent to each vertex, so that $N \geq 4N_0$, and we arrive at a contradiction.

In case not all of the edges are marked, we can follow the same lines. Let N_1' be the number of marked edges, N_0' the number of vertices at which the marked edges originate, and N_2' the number of regions into which the marked edges split up the surface of the polyhedron. By the lemma, a succession of marked edges cannot "terminate" by entering a vertex from which no other marked edge goes (this is why these edges split the polyhedron into regions). The network of marked edges need not be connected: they may form several components with no marked edges joining them. But the numbers a_k' of regions bounded by k marked edges and the numbers n_s' of marked edges originating from the sth vertex have the following two important properties: $a_2' = 0$ and $n_s' \geq 4$. Therefore the above scheme of allocating signs will again lead us to a contradiction if (instead of Euler's identity) we prove the inequality $N_0' - N_1' + N_2' \geq 2$.

In order to prove it, we will join successively the remaining edges to the marked ones, adding at each step an edge with at least one vertex belonging either to a marked edge or to an edge added before. After each step the number of edges increases by one, while the combined number of vertices and regions either increases by one or remains unchanged. Indeed, if at some step a new vertex appears, the number of regions does not change; and if the added edge connects two "old" vertices, then this edge either splits a region into two or splits up neither of the components.

Therefore at each step the quantity $\bar{N}_0 - \bar{N}_1 + \bar{N}_2$ either remains unchanged or decreases by one, and eventually, when all the edges have been added, we get $N_0 - N_1 + N_2 = 2$ by Euler's formula, which proves the inequality.

Thus all the bihedral angles of equivalent polyhedra are equal; hence when applying the two polyhedra to each other with a respective face, they will coincide as a whole. □

For the proof of Lemma 1 we need one more lemma.

LEMMA 2 (Cauchy–Steiner). *Suppose two convex polygons (planar or spherical) with vertices* $A_1 \cdots A_n$ *and* $A_1' \cdots A_n'$ *have equal lengths of the corresponding sides* $|A_1A_2| = |A_1'A_2'|, \dots, |A_{n-1}A_n| = |A_{n-1}'A_n'|$, *and their angles satisfy the inequalities* $\angle A_2 \leq \angle A_2', \dots, \angle A_{n-1} \leq \angle A_{n-1}'$. *Then* $|A_nA_1| \leq |A_n'A_1'|$.

PROOF. The lemma is obvious for triangles, because in triangles with two sides respectively equal the side opposite the larger angle is larger.

Turning to the general case, assume first that only one angle of the first polygon differs from the corresponding angle of the second, say $\angle A_i < \angle A_i'$, whereas $\angle A_j = \angle A_j'$, $j \neq i$, $i = 2, \ldots, n-1$. In this case the lemma is intuitively obvious: let us join the first and the last vertices with an elastic thread and put a hinge at a vertex distinct from these two. It is clear that increasing the angle at the hinged vertex (keeping the polygon otherwise rigid) will stretch the thread. Of course, this argument can be put in a precise form. Let us join A_i and A_i' with A_1, A_n and A_1', A_n' respectively. Then the polygons $A_i A_{i-1} \cdots A_1$ and $A_i' A_{i-1}' \cdots A_1'$, as well as the polygons $A_i A_{i+1} \cdots A_n$ and $A_i' A_{i+1}' \cdots A_n'$ are congruent, i.e., $|A_i A_n| = |A_i' A_n'|$ and $|A_i A_1| = |A_i' A_1'|$, so that we can again apply the theorem for triangles (since $\angle A_1 A_i A_n < \angle A_1' A_i' A_n'$).

If there are several angles of the first polygon that are less than the corresponding angles of the second, we will increase them successively, keeping the angles of the second polygon unchanged. At this point Cauchy finished the proof. The argument looked so obvious that no stumbling block could be expected.

However, there was a gap in Cauchy's argument, which was found by Steiner. The matter is that the above argument goes through provided the procedure of successively increasing the angles results each time in a convex figure. Unfortunately, this is not necessarily so. Nevertheless, the lemma is true, but Cauchy's arguments have to be somewhat refined (for details, see Lyusternik (1956)).

Lemma 2 immediately implies Lemma 1. Indeed, suppose there is a vertex such that the number of 1-edges adjacent to it is two. Let us intersect the surface of each polyhedron with a sphere centered at such vertex of a radius small enough to contain no other vertices. Then we obtain two spherical polygons with angles equal to the corresponding bihedral angles. If there is only one change from + to − between, say, A_1 and A_n and one change from − to + between, say, A_i and A_{i+1}, then we take some points B on $[A_1, A_n]$ and C on $[A_i, A_{i+1}]$ and the points B' and C' on the corresponding sides of the second polygon. Applying the Cauchy–Steiner lemma to the polygons $BA_1A_2 \cdots A_{i+1}C$ and $B'A_1'A_2' \cdots A_{i+1}'C'$, we obtain that $|BC| > |B'C'|$, and an application of this lemma to $BA_nA_{n-1} \cdots A_iC$ and $B'A_n'A_{n-1}' \cdots A_i'C'$ yields $|BC| < |B'C'|$. This contradiction proves Lemma 1. □

4.2. The Carathéodory, Radon, and Helly theorems.

The first two results in this section, though of independent interest, provide an important step towards the proof of Helly's theorem, which has numerous and fruitful applications in geometry and analysis.

THEOREM 1 (clean-up theorem for cones). *Let A be a nonempty subset of $\mathbb{R}^n$. Then any point $x \in \operatorname{cone} A$ (conic hull of A) different from the origin is representable as a conic combination of at most n points of A (i.e., there are a positive integer $r \le n$, points $x_i \in A$, and numbers $\mu_i > 0$, $1 \le i \le r$, such that $x = \sum_{i=1}^r \mu_i x_i$).*

PROOF. Let $x \in \operatorname{cone} A$ and $x \ne 0$. Then by definition $x = \sum_{i=1}^N \mu_i x_i$, $\mu_i > 0$. If $N \le n$, we have nothing to prove. Otherwise, if $N > n$, the vectors $\{x_i\}_{i=1}^N$ are linearly dependent; hence there are numbers γ_i, not all equal to zero, such that $\sum_{i=1}^N \gamma_i x_i = 0$. Without loss of generality we may assume that some of the numbers γ_i are positive (otherwise we reverse their sign). Let S be the set of the i's for which $\gamma_i > 0$. Set $\beta = \min_{i \in S} \frac{\mu_i}{\gamma_i}$ and $\mu_i' = \mu_i - \beta\gamma_i$, $1 \le i \le N$. Then all of the μ_i' are nonnegative with at least one of them equal to zero, and we have $\sum_{i=1}^N \mu_i' x_i = \sum_{i=1}^N \mu_i x_i - \beta \sum_{i=1}^N \gamma_i x_i = x$. Repeating this procedure we finally arrive at the required relation. □

COROLLARY (Carathéodory's theorem). *If A is a nonempty subset of $\mathbb{R}^n$, then any point of the convex hull of A is representable as a convex combination of at most $n+1$ points of A.*

PROOF. Consider the set $B = \{(1,x) \in \mathbb{R} \times \mathbb{R}^n \mid x \in A\}$. It is clear that $\operatorname{co} B = \{1\} \times \operatorname{co} A$. Let K be the cone generated by B. Then $\operatorname{co} B \subset K$. If $x \in \operatorname{co} A$, then $(1,x) \in \operatorname{co} B$ and by the clean-up theorem there are $r \le n+1$, points $(1, x_i) \in B$ (i.e., $x_i \in A$), and numbers $\mu_i > 0$, $1 \le i \le r$, such that $(1,x) = \sum_{i=1}^r \mu_i (1, x_i)$. Obviously, this is equivalent to $x \in \operatorname{co} \{x_i\}_{i=1}^r$, $x_i \in A$. □

THEOREM (Radon). *Any finite set in $\mathbb{R}^n$ consisting of at least $n+2$ points can be divided into two disjoint subsets such that the convex hulls of these subsets have a nonempty intersection.*

PROOF. Let $A = \{x_i\}_{i=1}^s$, $x_i \in \mathbb{R}^n$, and $s \ge n+2$. The vectors $\{x_i - x_1\}_{i=2}^s$ are linearly dependent; hence there are numbers μ_i,

$2 \le i \le s$, such that $0 = \sum_{i=2}^{s} \mu_i(x_i - x_1)$, i.e.,

$$\sum_{i=1}^{s} \nu_i x_i = 0, \quad \sum_{i=1}^{s} \nu_i = 0, \tag{i}$$

where $\nu_i = \mu_i$, $i \ge 2$, and $\nu_1 = \sum_{i=2}^{s} \mu_i$. Set $A_1 := \{x_i \in A \mid \nu_i \ge 0)\}$, $A_2 := \{x_i \in A \mid \nu_i < 0\}$, and

$$\overline{x} := a^{-1} \sum_{\{i \,|\, x_i \in A_1\}} \nu_i x_i, \quad \text{where} \quad a := \sum_{\{i \,|\, i \in A_1\}} \nu_i. \tag{ii}$$

Then

$$\overline{x} \overset{(ii)}{=} a^{-1} \sum_{\{i \,|\, x_i \in A_1\}} \nu_i x_i \overset{(i)}{=} a^{-1} \sum_{\{i \,|\, x_i \in A_2\}} (-\nu_i) x_i,$$

i.e., $\overline{x} \in \operatorname{co} A_1 \cap \operatorname{co} A_2$. □

THEOREM (Helly). *Let $\mathcal{A}$ be a set of indices and $\{A_\alpha\}_{\alpha \in \mathcal{A}}$ a family of closed convex subsets of $\mathbb{R}^n$ with at least one of them compact. If any subfamily consisting of $n+1$ sets has a nonempty intersection, then the whole family has a nonempty intersection.*

PROOF. First we prove by induction that any finite subfamily consisting, say, of s sets, $s \ge n+1$, has a nonempty intersection. Denote by $\operatorname{Ind}(s)$ the inductive statement of the sth step. By assumption $\operatorname{Ind}(n+1)$ is true. Assume that $\operatorname{Ind}(s-1)$ is true for some $s \ge n+2$ and prove $\operatorname{Ind}(s)$. Take any s indices $\{\alpha_i\}_{i=1}^{s}$. By $\operatorname{Ind}(s-1)$ there exists an $x_j \in \cap_{i=1, i \ne j}^{s} A_{\alpha_i}$, $j = 1, \dots, s$. According to Radon's theorem we can partition $\mathbb{N}_s := \{1, 2, \dots, s\}$ into disjoint subsets $\mathbb{N}'_s$ and $\mathbb{N}''_s$ such that there exists an $\overline{x} \in \operatorname{co}\{x_j\}_{j \in \mathbb{N}'_s} \cap \operatorname{co}\{x_j\}_{j \in \mathbb{N}''_s}$. But if $j \in \mathbb{N}''_s$, then $x_j \in A_{\alpha_i}$, $i \in \mathbb{N}'_s$. Hence $\overline{x} \in \cap_{i \in \mathbb{N}'_s} A_{\alpha_i}$. In a similar way, if $j \in \mathbb{N}'_s$, then $x_j \in A_{\alpha_i}$, $i \in \mathbb{N}''_s$, so that $\overline{x} \in \cap_{i \in \mathbb{N}''_s}$, i.e., $\overline{x} \in \cap_{i=1}^{s} A_{\alpha_i}$.

If $A_{\bar{\alpha}}$ is a compact set, then we have shown that the family of its closed subsets $\{A_\alpha \cap A_{\bar{\alpha}}\}_{\alpha \in \mathcal{A}}$ is centered (has the finite intersection property), which implies that the whole family has a nonempty intersection (see Dunford and Schwartz (1958)). □

4.3. Minkowski's theorem on the existence of a convex polyhedron.

Let M be a convex n-dimensional polyhedron, let $\{n_i\}_{i=1}^{k}$ be the unit normal vectors to its $(n-1)$-dimensional faces and $\{F_i\}_{i=1}^{k}$ the $(n-1)$-dimensional volumes of the faces. We will show that $\sum_{i=1}^{k} F_i n_i = 0$. Indeed, let Π be an arbitrary hyperplane and n the unit normal to it. Then $\langle n, n_i\rangle$ is the cosine of the angle between Π and the hyperplane passing through the face F_i, so that $\langle n, n_i\rangle F_i$ is the signed $(n-1)$-dimensional volume of the projection of F_i on Π. The projections on Π of the faces of the polyhedron M cover twice its projection on this hyperplane, and for each pair of points projecting into the same point the products $\langle n, n_i\rangle$ of the corresponding normals by n have opposite signs. Hence $\sum_{i=1}^k \langle n, n_i\rangle F_i = 0$, i.e., $\langle n, \sum_{i=1}^k F_i n_i\rangle = 0$ for any unit vector n. Therefore $\sum_{i=1}^k F_i n_i = 0$.

For a convex n-dimensional polyhedron the vectors $\{n_i\}_{i=1}^k$ must span the entire $\mathbb{R}^n$, since otherwise they would lie in a proper subspace so that the polyhedron would be unbounded in a normal direction to this subspace.

Minkowski proved that the above conditions on the vectors n_i and the numbers F_i are not only necessary but also sufficient for the existence of the corresponding polyhedron.

THEOREM (Minkowski). *Let $\{n_i\}_{i=1}^k$ be unit vectors in $\mathbb{R}^n$ that span the entire $\mathbb{R}^n$, and $\{F_i\}_{i=1}^k$ positive numbers such that $\sum_{i=1}^k F_i n_i = 0$. Then there exists a convex n-dimensional polyhedron with volumes of its $(n-1)$-dimensional faces equal to F_i and normal vectors to its faces equal to n_i.*

PROOF. The existence of the required polyhedron follows from the existence of a solution to the following extremal problem:

$$\begin{aligned} &\mathrm{Vol}_n\{x \in \mathbb{R}^n \mid \langle x, n_i\rangle \le h_i,\ 1 \le i \le k\} \to \max, \\ &\sum_{i=1}^k F_i h_i = 1, \quad h_i \ge 0, \qquad 1 \le i \le k. \end{aligned} \tag{P}$$

For positive h_i the set in braces is, obviously, a convex polyhedron $M(h)$, $h = (h_1, \dots, h_k)$, of dimension k, since this set contains the ball $\{x \in \mathbb{R}^n \mid |x| \le \min_{1\le i\le k} h_i\}$.

Denote by $V(h)$, $h = (h_1, \dots, h_n)$, the functional to be maximized in the problem (P). The function $h \to V(h)$ is continuous; hence (by the Weierstrass theorem) it attains maximum on the $(k-1)$-dimensional simplex

$$\Big\{h = (h_1, \dots, h_k) \in \mathbb{R}^n \mid \sum_{i=1}^k F_i h_i = 1,\ h_i \ge 0,\ i = 1, \dots, k\Big\}$$

at some point $\widehat{h} = (\widehat{h}_1, \dots, \widehat{h}_k)$, $i = 1, \dots, k$. Note also that in a neighborhood of $\widehat{h}$ the function $V(\cdot)$ is differentiable and $(\partial V/\partial h_i)(\widehat{h}) = F_i(\widehat{h})$, where $F_i(\widehat{h})$ is the $(n-1)$-dimensional volume of the intersection of the polyhedron $M(\widehat{h})$ with the hyperplane $\{x \in \mathbb{R}^n \mid \langle x, n_i\rangle = \widehat{h}_i\}$. (Indeed, if $F_i(\widehat{h}_i) > 0$, then substituting $\widehat{h}_i + t$ for $\widehat{h}_i$ we add to $M(\widehat{h})$ (for $t > 0$) or subtract from $M(\widehat{h})$ (for $t < 0$) a figure of volume $tF_i(\widehat{h}_i)$ up to higher order terms for small t, or of volume $o(t)$ if $F_i(\widehat{h}) = 0$.)

Applying the Lagrange principle we obtain that $F_i(\widehat{h}) = \lambda F_i$ and

$$1 = \sum_{i=1}^{k} F_i(\widehat{h})\widehat{h}_i = \lambda \sum_{i=1}^{k} F_i \widehat{h}_i = \lambda.$$

This shows that the ith face of the polyhedron $M(\widehat{h})$ is of volume F_i. □

4.4. The Cauchy and Steiner–Minkowski formulas.

THEOREM (Cauchy's formula). *Let $M \subset \mathbb{R}^n$ be a convex polyhedron, let $d\mu$ be the measure on the sphere $\mathbb{S}^{n-1}$ invariant with respect to translations, $\int_{\mathbb{S}^{n-1}} d\mu = \operatorname{meas} \mathbb{S}^{n-1}$, and let $V_N(\xi)$ be the $(n-1)$-dimensional volume of the projection of M on the hyperplane orthogonal to the unit vector $\xi \in \mathbb{S}^{n-1}$. Then*

$$\int_{\mathbb{S}^{n-1}} V_M(\xi)\, d\mu = \beta_{n-1} V_{\partial M}, \tag{1}$$

where $V_{\partial M}$ is the volume of the boundary of M; i.e., the sum of volumes of its faces, and β_{n-1} is the volume of the $(n-1)$-dimensional unit ball.

PROOF. Let Γ_i be one of the faces of the polyhedron M and $V_{\Gamma_i}(\xi)$ the volume of its projection on the hyperplane orthogonal to the unit vector $\xi \in \mathbb{S}^{n-1}$. It is clear that $V_{\Gamma_i}(\xi) = |\langle \xi, n_i\rangle| V_{\Gamma_i}$, where n_i is the unit normal to the face Γ_i and V_{Γ_i} is the volume of this face. Invariance of the measure $d\mu$ implies that $\int_{\mathbb{S}^{n-1}} |\langle \xi, a\rangle| d\mu$ does not depend on the choice of a and hence is equal to some number c_n dependent only on n. Therefore

$$\int_{\mathbb{S}^{n-1}} V_{\Gamma_i}(\xi) d\mu = c_n V_{\Gamma_i}.$$

Each point of the projection of M (except a set of null measure) is covered by projections of *two* faces. This implies that

$$\int_{\mathbb{S}^{n-1}} V_M(\xi)d\mu = \int_{\mathbb{S}^{n-1}} \frac{1}{2}\sum_i V_{\Gamma_i}(\xi)d\mu = \frac{c_n}{2}\sum_i V_{\Gamma_i} = \frac{c_n}{2}V_{\partial M}.$$

To evaluate the constant c_n, it is expedient to consider the unit ball B^n instead of M (treating $\mathbb{S}^{n-1}$ as a limit of polyhedra inscribed in $\mathbb{S}^{n-1}$). In this case $V_{B^n}(\xi) = \beta_{n-1}$, so that $c_n/2 = \beta_{n-1}$. □

REMARK. Cauchy's formula implies that if one convex polyhedron lies inside another, the volume of the inner polyhedron does not exceed that of the outer. By approximating a convex body by convex polyhedra and taking the limit, it is not hard to establish Cauchy's formula for an arbitrary convex body M.

Let $M \subset \mathbb{R}^n$ be a convex polyhedron and M_r, $r \geq 0$, the set of points in $\mathbb{R}^n$ distant at most by r from M (in Euclidean norm); i.e., M_r is the union of all balls of radius r with centers in M. One can easily verify that M is a convex set.

THEOREM (The Steiner–Minkowski formula). *The volume V_r of the set M_r depends on r polynomially; i.e.,*

$$V_r = \beta_0 + \beta_1 r + \cdots + \beta_n r^n, \tag{2}$$

where $\beta_0 = V_0$ is the volume of M, β_1 is the volume of its surface, and β_n is the volume of the n-dimensional ball.

PROOF. We will conduct the proof by induction in n using Cauchy's formula for convex bodies (see the remark above). For $n = 1$ formula (2) is obvious. Assume that is true for $(n-1)$-dimensional polyhedra and let M be a convex polyhedron in $\mathbb{R}^n$. Denote by $P_M(\xi)$ the projection of M on the hyperplane orthogonal to a unit vector ξ. It is clear that $P_{M_r}(\xi) = (P_M(\xi))_r$. Let $V_r(\xi)$ be the $(n-1)$-dimensional volume of $P_{M_r}(\xi)$. By the inductive assumption, $V_r(\xi) = V_0(\xi) + \cdots + \beta_{n-1}r^{n-1}$. According to Cauchy's formula

$$\begin{aligned}\beta_{n-1}V_{\partial M_r} &= \int_{\mathbb{S}^{n-1}} V_{M_r}(\xi)d\mu = \int_{\mathbb{S}^{n-1}} \sum_{k=0}^{n-1} \beta_k (P_{M_r}(\xi))_r d\mu \\ &= \beta_0 V_0 n - 1 + \sum_{k=1}^{n-1} a_k r^k,\end{aligned}$$

where the a_k are some constants.

It is clear from geometrical considerations that $dV_r/dr = P_r$, where P_r is the volume of the surface of M_r, so that

$$V_r = V_0 + \int_0^r P_t dt = V_0 + \int_0^r \left(P_0 + \sum_{k=1}^{n-1} a'_k t^k\right) dt$$
$$= V_0 + P_0 r + \sum_{k=1}^{n-1} b_k r^{k+1}.$$

□

4.5. The Brunn–Minkowski inequality and the Grünbaum–Hammer theorem.

The Grünbaum–Hammer theorem relies on the Brunn–Minkowski inequality and symmetrization. Both these facts are of independent interest.

Let $A \subset \mathbb{R}^n$. Denote by $\Gamma_{y,a}$ the hyperplane $\{x \in \mathbb{R}^n \mid \langle x, y\rangle = a\}$. Writing $\mathrm{Vol}_{n-1}(C)$ for the $(n-1)$-dimensional volume of an $(n-1)$-dimensional set C, set

$$\varphi_y(a) = \begin{cases} \mathrm{Vol}_{n-1}^{1/(n-1)}(A \cap \Gamma_{y,a}), & (A \cap \Gamma_{y,a}) \neq \emptyset, \\ -\infty, & (A \cap \Gamma_{y,a}) = \emptyset. \end{cases}$$

THEOREM (The Brunn–Minkowski inequality). *Let A be a convex compact set in $\mathbb{R}^n$. Then for any $y \in \mathbb{R}^n \setminus \{0\}$ and $-\infty < a_0 < a_1 < \infty$, the following inequality holds:*

$$\varphi_y(\alpha a_0 + (1-\alpha)a_1) \geq \alpha\varphi_y(a_0) + (1-\alpha)\varphi_y(a_1).$$

Let A be a convex compact body in $\mathbb{R}^n$, and let $e = (1, 0, ..., 0)$, $M := \max\{\langle e, x\rangle \mid x \in A\}$, $m := \min\{\langle e, x\rangle \mid x \in A\}$. Set

$$\sigma(A) := \{x \in \mathbb{R}^n \mid x = (t, \xi),\ t \in [m, M],\ |\xi| \leq r(t)\},$$

where $r(t)$ is such that $\mathrm{Vol}_{n-1} B(0, r(t), \mathbb{R}^{n-1}) = \mathrm{Vol}_{n-1}(A \cap \Gamma_{e,t})$. The set $\sigma(A)$ is the *Brunn–Minkowski symmetrization* of A.

It follows from the Brunn–Minkowski inequality that the Brunn–Minkowski symmetrization $\sigma(A)$ is a convex compact set (provided so is A itself) and

$$\mathrm{Vol}_n(A) = \mathrm{Vol}_n(\sigma(A)). \qquad (*)$$

For a set $C \in \mathbb{R}^n$ denote by $\operatorname{gr} C$ its center of gravity. If $x_* = (x_{*1}, \dots, x_{*n}) = \operatorname{gr} A$, then the theorems of integral calculus imply that

$$\operatorname{gr} \sigma(A) = (x_{*1}, 0, \dots, 0). \tag{**}$$

THEOREM (Grünbaum–Hammer). *Let A be a compact convex body in $\mathbb{R}^n$ and x_* its center of gravity. Then any hyperplane $\Gamma_{x_*, \langle y, x_* \rangle}$, $y \neq 0$, cuts off from A a volume whose fraction relative to that of A is at least $(n/(n+1))^n$ $(\geq 1/e)$; i.e.,*

$$\operatorname{Vol}_n(A \cap \Pi_{y, \langle y, x_* \rangle}) \geq (n/(n+1))^n \operatorname{Vol}_n(A), \quad \forall y \in \mathbb{R}^n, \ y \neq 0.$$

4.6. Some complements.

Here we briefly review some more results from the theory of convex bodies and polyhedra, which is one of the earliest and most beautiful branches of geometry. It goes back to the theory of Platonic solids and Cauchy's theorem on polyhedra. The basic subject of this theory is compact convex sets in $\mathbb{R}^n$, which are also referred to as *convex figures*.

Define a distance in the set of convex figures by setting

$$h(A_1, A_2) = \inf\{\varepsilon \geq 0 \mid A_1 + \varepsilon B \supset A_2, \ A_2 + \varepsilon B \supset A_1\},$$

where B is the unit ball in $\mathbb{R}^n$. The quantity $h(A_1, A_2)$ is called the *Hausdorff distance* between A_1 and A_2. Note also that since convex sets can be added and multiplied by numbers,

$$A_1 + A_2 = \{x \mid x = x_1 + x_2, \ x_1 \in A_1, x_2 \in A_2\},$$
$$\alpha A = \{x \mid x = \alpha y, \ y \in A\},$$

the set of all convex figures forms a cone. This cone is embedded into the "linear space of convex figures" as a semigroup is embedded into a group.

We state the following important theorem.

THEOREM (Blaschke's compactness theorem). *The set of convex figures forms a locally compact cone (in topology induced by the Hausdorff metric).*

In many cases this theorem allows for proving theorems on existence of solutions to extremal problems concerning convex figures.

The convex hull of a finite number of points is a *convex polyhedron*.

THEOREM (on approximation). *Let A be a convex figure in $\mathbb{R}^n$ and $\varepsilon > 0$. Then there are two convex polyhedra A_- and A_+ such that $A_- \subset A \subset A_+$ and $h(A_-, A_+) < \varepsilon$; i.e., a convex compact set may be approximated by polyhedra with any accuracy.*

This theorem enables one to carry over many facts of the geometry of polyhedra to convex surfaces.

The study of convex polyhedra began in ancient times. The 13th book of Euclid's *Elements* is devoted to five regular convex polyhedra in $\mathbb{R}^3$, the Platonic solids: regular tetrahedron, cube, octahedron, dodecahedron, and icosahedron.

In $\mathbb{R}^4$ there are six regular convex polyhedra, and in $\mathbb{R}^n$ for $n \geq 5$ only three: regular simplex, cube, and octahedron.

An important role in various fields of natural science is played by regular convex polyhedra which regularly partition the space, but this subject is beyond the scope of this book.

An intersection of finitely many half-spaces forms a *polyhedral set.*

THEOREM (on polyhedral sets). *A set P is polyhedral if and only if it is finitely generated, i.e., if there exist $\xi_i \in \mathbb{R}^n$, $i = 1, \dots, m$, and a positive integer k, $0 \leq k \leq m$, such that for any $x \in P$ there are numbers $\lambda_i \geq 0$, $i = 1, \dots, m$, such that $x = \sum_{i=1}^k \lambda_i \xi_i + \sum_{i=k+1}^m \lambda_i \xi_i$, $\sum_{i=1}^k \lambda_i = 1$.*

This implies, in particular, that a bounded polyhedral set is a convex polyhedron and a finitely generated cone is closed because it is polyhedral.

5. Convex extremal problems

- In this section we discuss conditions for extremum and duality in convex extremal problems.

5.1. Conditions for a minimum in extremal problems.

Let $f\colon \mathbb{R}^n \to \overline{\mathbb{R}}$ be a convex function and C a nonempty convex subset of $\mathbb{R}^n$. The minimization problem of f over the set C (or, more precisely, the problem of finding and/or characterizing the points in C where the restriction of f to C attains its minimum) will be formally written as

$$f(x) \to \min, \quad x \in C, \tag{P}$$

and referred to as a *convex problem.*

The set C is the *set of constraints* or simply the *constraint* in the problem (P). If $C = \mathbb{R}^n$, then (P) is a *problem without constraints*. The points of C are said to be *admissible* (in the problem (P)).

An admissible point $\widehat{x}$ is a *solution* to the problem (P) or a *minimum point* in the problem (P) if $f(x) \geq f(\widehat{x})$ for all $x \in C$. The minimal value of f in problem (P) is called the *value* of this problem.

In a problem without constraints, necessary and sufficient conditions for $\widehat{x}$ to be a minimum point are given by an analog of the Fermat theorem stated in Section 3.

In this section we will be interested in conditions for a minimum in a convex problem when C is not the entire $\mathbb{R}^n$ and is specified by certain relations such as equalities, inequalities, and inclusions.

Suppose we are given a nonempty convex subset A of $\mathbb{R}^n$, a convex proper function $f_0 \colon \mathbb{R}^n \to \overline{\mathbb{R}}$ finite on A, convex functions $f_i \colon \mathbb{R}^n \to \mathbb{R}$, $i = 0, 1, \dots, m'$, and affine functions $f_i \colon \mathbb{R}^n \to \mathbb{R}$, $i = m' + 1, \dots, m$ (recall that an affine function is the sum of a linear function and a constant). The problem

$$f_0(x) \to \min,$$
$$f_i(x) \leq 0,\ i = 1, \dots, m', \quad f_i(x) = 0,\ i = m' + 1, \dots, m, \quad x \in A, \tag{P_1}$$

is obviously a convex one. It is usually referred to as a *convex programming problem*.

The function $\mathcal{L} \colon \mathbb{R}^n \times \mathbb{R}^{m+1} \to \overline{\mathbb{R}}$, $\mathcal{L}(x, \lambda) = \sum_{i=0}^{m} \lambda_i f_i(x)$, where $\lambda = (\lambda_0, \lambda_1, \dots, \lambda_m)$, is called the *Lagrange function of the problem* (P_1) and the numbers λ_i, $i = 0, 1, \dots, m$, are the *Lagrange multipliers*; the vector λ will be referred to as the set of Lagrange multipliers.

THEOREM (Karush–Kuhn–Tucker). *Let $\widehat{x}$ be a solution to the problem (P_1). Then there is a set of Lagrange multipliers $\widehat{\lambda} = (\widehat{\lambda}_0, \widehat{\lambda}_1, \dots, \widehat{\lambda}_m)$, not all of which vanish, such that*

(a) $\widehat{\lambda}_i \geq 0$, $i = 0, 1, \dots, m'$;

(b) $\widehat{\lambda}_i f(\widehat{x}) = 0$, $i = 1, \dots, m'$;

(c) $\min_{x \in A} \mathcal{L}(x, \widehat{\lambda}) = \mathcal{L}(\widehat{x}, \widehat{\lambda})$.

If a point $\widehat{x}$ admissible in the problem (P) satisfies conditions (a), (b), and (c) with $\widehat{\lambda}_0 > 0$, then $\widehat{x}$ is a solution to the problem (P).

If $m' = m$, $\widehat{x}$ is a solution to the problem (P) and there is a point $\overline{x}$ admissible in this problem such that $f_i(\overline{x}) < 0$, $i = 1, \dots, m$, then $\widehat{\lambda}_0 \neq 0$ (Slater's condition).

The equalities (b) in the theorem are called the *complementary slackness* conditions.

PROOF. Let $\widehat{x}$ be a solution to the problem (P_1). Consider the set

$$B = \{b = (b_0, b_1, \dots, b_m)^T \in \mathbb{R}^{m+1} \mid \exists\, x \in A\colon \\ f_i(x) \leq b_i,\ 0 \leq i \leq m';\ f_i(x) = b_i,\ m'+1 \leq i \leq m\}.$$

Since $\widehat{b} = (f_0(\widehat{x}), 0, \dots, 0)^T \in B$ for $x = \widehat{x}$, we have that $B \neq \emptyset$, and it can be easily verified that B is a convex set. Then $\operatorname{ri} B \neq \emptyset$ by Proposition 1 of Section 1. We will show that $\widehat{b} \notin \operatorname{ri} B$. Indeed, otherwise there is an $\varepsilon > 0$ such that $\widehat{b} + L_B \cap \dot{B}_\varepsilon(0) \subset B$, where L_B is the subspace from which $\operatorname{aff} B$ is obtained by translation. Since $(f_0(\widehat{x}) + 1, 0, \dots, 0)^T \in B$, we have $(-1, 0, \dots, 0)^T = (f_0(\widehat{x}), 0, \dots, 0)^T - (f_0(\widehat{x}) + 1, 0, \dots, 0)^T \in L_B$, and since $(\varepsilon/2)(-1, 0, \dots, 0) \in \dot{B}_\varepsilon(0)$, we have $\widehat{b} + (\varepsilon/2)(-1, 0, \dots, 0)^T = (f_0(\widehat{x}) - \varepsilon/2, 0, \dots, 0)^T \in B$. But this implies that there is an admissible point $\overline{x}$ such that $f_0(\overline{x}) \leq f_0(\widehat{x}) - \varepsilon/2 < f_0(\widehat{x})$. This contradicts the assumption that $\widehat{x}$ is a minimum point in the problem (P_1); therefore $\widehat{b} \notin \operatorname{ri} B$.

By the first separation theorem (see Section 1) there exists a nonzero vector $\widehat{\lambda} = (\widehat{\lambda}_0, \widehat{\lambda}_1, \dots, \widehat{\lambda}_m)$ such that

$$\sum_{i=0}^{m} \widehat{\lambda}_i b_i \geq \widehat{\lambda}_0 f_0(\widehat{x}) \qquad \forall\, b = (b_0, b_1, \dots, b_m) \in B. \tag{i}$$

Substituting for b the vectors

$$(f_0(\widehat{x}) + 1, 0, \dots, 0)^T, (f_0(\widehat{x}), 1, 0, \dots, 0)^T, \dots, \\ (f_0(\widehat{x}), 0, \dots, 0, 1, 0, \dots, 0)^T,$$

where the unity in the last vector stands on the m'th place, we obtain that $\widehat{\lambda}_i \geq 0$, $i = 0, 1, \dots, m'$, which proves (a).

Now putting into (i) the vectors

$$(f_0(\widehat{x}), f_1(\widehat{x}), 0, \dots, 0)^T, \dots, (f_0(\widehat{x}), 0, \dots, 0, f_{m'}(\widehat{x}), 0, \dots, 0)^T,$$

we obtain that $\widehat{\lambda}_i f_i(\widehat{x}) \geq 0$, $i = 1, \dots, m'$. But $\widehat{\lambda}_i f_i(\widehat{x}) \leq 0$, $i = 1, \dots, m'$, because of (a) and because $\widehat{x}$ is an admissible point. Thus $\widehat{\lambda}_i f_i(\widehat{x}) = 0$, $i = 1, \dots, m'$, which proves (b).

Let x be an admissible point of (P_1). Then $(f_0(x), f_1(x), \dots, f_m(x))^T \in B$, and we have

$$\mathcal{L}(x, \widehat{\lambda}) = \sum_{i=0}^{m} \widehat{\lambda}_i f_i(x) \overset{(i)}{\geq} \widehat{\lambda}_0 f_0(\widehat{x}) \overset{(b)}{=} \sum_{i=0}^{m} \widehat{\lambda}_i f_i(\widehat{x}) = \mathcal{L}(\widehat{x}, \widehat{\lambda}),$$

where the second-to-last equality uses $f_i(\widehat{x}) = 0$, $m' + 1 \leq i \leq m$. This proves (c).

Let conditions $(a) - (c)$ be fulfilled with $\widehat{\lambda}_0 > 0$. Then for any admissible point x we have

$$\widehat{\lambda}_0 f_0(x) \geq \widehat{\lambda}_0 f_0(x) + \sum_{i=1}^{m} \widehat{\lambda}_i f_i(x) \overset{(c)}{\geq} \widehat{\lambda}_0 f_0(\widehat{x}) + \sum_{i=1}^{m} \widehat{\lambda}_i f_i(\widehat{x}) \overset{(b)}{=} \widehat{\lambda}_0 f_0(\widehat{x});$$

i.e., $\widehat{x}$ is a minimum point in the problem (P_1).

Let us prove the last statement. Suppose that $\widehat{\lambda}_0 = 0$. Then some of the $\widehat{\lambda}_i$ are positive, and we have $\sum_{i=0}^{m} \widehat{\lambda}_i f(\overline{x}) = \sum_{i=1}^{m} \widehat{\lambda}_i f(\overline{x}) < 0 = \sum_{i=0}^{m} \widehat{\lambda}_i f(\widehat{x})$, which is impossible. □

REMARK. The Karush–Kuhn–Tucker theorem remains valid if we consider an arbitrary vector space X instead of $\mathbb{R}^n$. Then the proof of necessity has to be modified, but the proof of sufficiency, obviously, does not depend on the structure of X (this will be used in Section 10).

5.2. Duality of convex problems.

We begin with a general scheme of constructing a dual problem to a given one. Consider the problem

$$f(x) \to \min, \quad x \in \mathbb{R}^n. \tag{1}$$

We will embed this problem into a family of "similar" problems depending on some parameter (or, as is said sometimes, we *perturb* the problem). More precisely, let $F\colon \mathbb{R}^m \times Y \to \overline{\mathbb{R}}$ be a function such that $F(x, 0) = f(x)$ for any $x \in \mathbb{R}^n$. For each $y \in \mathbb{R}^m$ we consider the problem

$$F(x, y) \to \min, \quad x \in \mathbb{R}^n. \tag{1_y}$$

The family of such problems is called a *perturbation of the problem* (1), and the function $S\colon \mathbb{R}^m \to \overline{\mathbb{R}}$ specifying for each y the value $S(y)$ of the problem (1_y) is called the *S-function of the family* (1_y). Clearly, $S(0)$ is the value of the initial problem (1).

Before defining the dual problem to (1) we present the reasons motivating this definition. If F is convex, then it is not hard to check that the S-function is convex as well; and if the latter is closed, then by the Fenchel–Moreau theorem $S = S^{**}$ and, in particular, $S(0) = S^{**}(0)$. Let us describe the problem having the value $S^{**}(0)$, which will be taken for the dual to (1). By definitions $S(0) \geq S^{**}(0) = \sup_{y' \in (\mathbb{R}^m)'} (-S^*(y'))$ and

$$\begin{aligned} S^*(y') &= \sup_{y \in \mathbb{R}^m} \Big(y' \cdot y - \inf_{x \in \mathbb{R}^n} F(x, y) \Big) \\ &= \sup_{x \in \mathbb{R}^n, y \in \mathbb{R}^m} (0 \cdot x + y' \cdot y - F(x, y)), \end{aligned}$$

where the last expression gives the function conjugate with F at the point $(0, y')$.

Hence the *dual problem to* (1) (with respect to a given perturbation) is the problem

$$-F^*(0, y') \to \max, \quad y' \in (\mathbb{R}^m)'. \tag{1*}$$

Since the function $F^*(0, \cdot)$ is convex, we see that (1^*) is the maximization problem of a concave function. We call such problems convex as well (since instead of (1^*) one can consider the minimization problem of $F^*(0, \cdot)$; this replacement changes the sign of the value and does not affect the solution).

Thus the dual problem is convex regardless of whether or not the initial problem is convex. The above relations imply that the *value* of the dual problem (1^*) is always no greater than that of the initial problem (1) and that a sufficient condition for their values to be equal is $S(0) = S^{**}(0)$. The conditions to ensure this equality are, of course, based on the Fenchel–Moreau theorem.

Let us write down the dual problem to (P) (assuming for simplicity that $m' = m$; i.e., there are no constraints of equality type) with respect to the perturbation

$$f_0(x) \to \min, \quad f_i(x) \leq y_i, \ i = 1, \dots, m, \quad x \in A,$$

which will be referred to as the *standard* perturbation. In terms of the general scheme this means that we write the initial problem in the form

$$f(x) \to \min, \quad x \in \mathbb{R}^n,$$

where $f(x) = f_0(x)$ if $f_i(x) \leq 0$, $i = 1, \dots, m$, and $x \in A$, and $f(x) = +\infty$ otherwise, and then we consider the function $F \colon \mathbb{R}^n \times$

$\mathbb{R}^m \to \overline{\mathbb{R}}$ such that $F(x,y) = f_0(x)$ if $f_i(x) \le y_i$, $i = 1, \dots, m$ (where $y = (y_1, \dots, y_m)^T$) and $x \in A$, and $F(x,y) = +\infty$ otherwise.

Let us find the function conjugate with F at the point $(0, y')$:

$$\begin{aligned} F^*(0,y') &= \sup_{x\in\mathbb{R}^n, y\in\mathbb{R}^m} (y'\cdot y - F(x,y)) \\ &= \sup_{x\in A} \sup_{\substack{y\in\mathbb{R}^m \\ f_i(x)\le y_i, i=1,\dots,m}} (y'\cdot y - f_0(x)). \end{aligned}$$

If $y' = (y'_1, \dots, y'_m) \ge 0$, i.e., all the coordinates are nonnegative, then the second supremum is easily seen to equal $+\infty$; and if $y' \le 0$, then it is equal to $\sum_{i=1}^m f_i(x)y'_i - f_0(x)$. Thus the dual problem to (P) (with respect to the standard perturbation) has the form

$$-\sup_{x\in A}\left(\sum_{i=1}^m f_i(x)y'_i - f_0(x)\right) \to \max, \quad y'_i \le 0, \ i = 1, \dots, m. \quad (P^*)$$

Let us apply the formulas just obtained to the *linear programming problem.* This is a particular case of a convex programming problem which consists in maximization of a linear functional over a convex polyhedron. The fundamentals of the theory of such problems were laid by L. V. Kantorovich in the late thirties. His results were rediscovered by American mathematicians (Kuhn, Tucker, Dantzig, and others) during the 40s.

The linear programming problem may be written in different forms. We will consider this problem in the so-called *normal form*:

$$c\cdot x \to \min, \qquad Ax \ge b, \quad x \ge 0, \tag{2}$$

where, as usual, $x \in \mathbb{R}^n$, $c \in (\mathbb{R}^n)'$, A is an $m\times n$ matrix, $b \in \mathbb{R}^m$, and inequalities between vectors are to be understood coordinate-wise.

Writing $f_0(x) = c\cdot x$, $f_i(x) = \langle -a^i, x\rangle + b_i$, $i = 1, \dots, m$, where the a^i are the rows of the matrix A and the b_i are the coordinates of the vector b, and letting $A = \mathbb{R}^n_+$, we see that (2) is a particular case of the problem (P_1).

Due to the fairly simple structure of the linear programming problem, its dual problem (with respect to the standard perturbation) can be written in an explicit form, and it is also a linear programming problem. Indeed, the perturbation (1) is written as

$$c\cdot x \to \min, \quad -Ax + b \le y, \quad x \ge 0.$$

Then (see problem (P^*))

$$-F^*(0,y') = -\sup_{x\geq 0}(y'\cdot(-Ax+b)-c\cdot x) = \inf_{x\geq 0}(-y'\cdot b+(y'A+c)\cdot x).$$

The set of those y' where this function equals $-\infty$ can be discarded because we are interested in its supremum. Thus $y'A \geq -c$ and in this case $-F^*(0,y') = -y'\cdot b$. Introducing the variable $\xi = -y'$, the dual problem to (1) is written as

$$\xi\cdot b \to \max, \quad \xi A \leq c, \quad \xi \geq 0.$$

5.3. Algorithms of convex optimization.

The algorithms for finding the solutions to extremal problems are based on the ideas of the *method of descent* as well as on the *penalization* and *central section* methods. In the infinite-dimensional case there are various methods of reduction to a finite-dimensional setup.

The method of central sections. Consider the minimization problem of a *convex differentiable function f over a convex finite-dimensional compact body $A \subset \mathbb{R}^d$*:

$$f(x) \to \min, \quad x \in A. \tag{P_2}$$

This is a general problem of convex finite-dimensional optimization.

In the mid-60s, A. Levin in Russia and D. Newman in the USA proposed the following method of seeking a minimum in problem (P_2) based on the Grünbaum–Hammer theorem, which was proved in the previous section. According to this theorem a hyperplane passing through the center of gravity of a convex body A in d-dimensional space splits the body into two sets, A' and A'', with the volume of each of them no less than $(1-e^{-1})$ times the volume of the entire set A.

The method itself (referred to as the *method of central sections*) consists of the following procedure. Let us write A_0 for A. Find its center of gravity $x_1 = \operatorname{gr} A_0$ and compute $f'(x_1)$. If it equals the zero vector, x_1 is the minimum point and the problem is solved. Otherwise, we may discard the part of A_0 lying in the half-space $\Pi_0' := \{x \mid f'(x_1)\cdot(x-x_1) > 0\}$ (since, as we remember, any convex smooth function fulfills the inequality $f(x)-f(\xi) \geq f'(\xi)\cdot(x-\xi)$, so that for $x \in A_0 \cap \Pi_1'$ we will have $f(x) > f(x_1) \geq \min f$). Denote by A_1 the remaining part after deleting $A \cap \Pi_0'$ (where $A = A_0$) and implement the procedure repeatedly.

Now at the nth step we select the point ξ_n among $\{x_1, ..., x_n\}$ where the value of f is no greater than any of the values $\{f(x_i), 1 \leq i \leq n\}$. We will prove that $f(\xi_n)$ converges to the value of the problem (P_2) at a geometric rate. Indeed, without loss of generality we may assume that $0 \in A$ and that this is the minimum point in problem (P_2). Let $\alpha > (\operatorname{Vol}_d A_n / \operatorname{Vol}_d A)^{1/d} \geq \alpha/2$ (where $\operatorname{Vol}_d C$ is the volume of the d-dimensional set C). Then by definition $\operatorname{Vol}_d(\alpha A) > \operatorname{Vol}_d A_n$; hence there is an element $\overline{x} \in \alpha A \backslash A_n$. We see from the construction of the algorithm that if the element $\overline{x}$ was discarded, then $f(\overline{x}) > f(x_s)$ for some s; hence $f(\overline{x}) > f(\xi_n)$. But $\overline{x} \in \alpha A$, so that $\overline{x} = \alpha\xi$, $\xi \in A$; therefore by the Grünbaum–Hammer theorem we have (denoting by $\operatorname{var} f$ the maximum of $f(x) - f(0)$, $x \in A$):

$$\begin{aligned} f(\xi_n) &< f(\overline{x}) = f(\alpha\xi + 0(1-\alpha)) \\ &\leq \alpha f(\xi) + (1-\alpha) f(0) = f(0) + \alpha(f(\xi) - f(0)) \\ &\leq f(0) + \alpha \operatorname{var} f \overset{\operatorname{def}\alpha}{\leq} f(0) + 2(\operatorname{Vol}_d A_n / \operatorname{Vol}_d A)^{1/d} \operatorname{var} f \\ &\leq f(0) + (1 - e^{-1})^{n/d} \operatorname{var} f. \qquad \square \end{aligned}$$

The method of central sections has not received practical implementation because of computational difficulties in finding the center of gravity. But its basic idea can be used for constructing algorithms of much practical use. Here is one such algorithm.

The method of circumscribed ellipsoids. This method is based on a combination of two ideas, viz., the idea of section discussed above and the following geometric fact: *half of an ellipsoid can be placed inside an ellipsoid of smaller volume than that of the initial ellipsoid*! It is essential that, first, the ratio of the volume of the ellipsoid circumscribing the "semi-ellipsoid" to that of the ellipsoid itself is less than one and, secondly, the computation of the center of the new ellipsoid from the position of the semi-ellipsoid requires order d^2 operations. Let us show how the circumscribed ellipsoid is constructed.

Due to the affine nature of the problem (i.e., because the ratio of volumes of geometric fugures does not change under their affine transformations) we may assume that the initial ellipsoid is the ball $\sum_{k=1}^d x_k^2 \leq 1$ and the semi-ellipsoid is the "upper hemisphere", i.e., consists of the points of the ball with nonnegative last coordinate. Let us take the point ξ with coordinates $(0, \ldots, 0, 1/(d+1))$ for the center of the circumscribed ellipsoid and define the ellipsoid by the inequalities

$$\sum_{k=1}^{d-1} \frac{x_k^2(d^2-1)}{d^2} + \frac{(x_d - \frac{1}{d+1})^2(d+1)^2}{d^2} \le 1.$$

The volume of this ellipsoid equals the product of its semi-axes times the volume of the unit ball. But we are interested in its ratio to the latter, which is therefore equal simply to the product of semi-axes:

$$g(d) = \left(\frac{d}{\sqrt{d^2-1}}\right)^{d-1} \frac{d}{d+1} = \frac{d^d}{(d-1)^{(d-1)/2}(d+1)^{(d+1)/2}}.$$

It is not hard to prove (as a consequence of convexity of the function $d \to d \log d$) that $g(d) < 1$ for any positive integer $d \ge 2$.

Now we can describe the algorithm. Let E_0 be an ellipsoid containing the set A. If its center c_0 does not belong to A, we can construct a hyperplane passing through c_0 and containing no points of A. Then we discard the half of the ellipsoid disjoint with A. Otherwise, if $c_0 \in A$, we compute $f'(c_0)$ and perform the central section "according to Levin–Newman" to obtain again a semi-ellipsoid, which we denote by E_0'. Now we circumscribe around E_0' an ellipsoid of smaller volume than that of E_0, denote this ellipsoid by E_1, and iterate the procedure. Again, we will approach the value of the problem at a geometric rate.

The simplex method for the solution of linear programming problems. We will describe the method for finding the solutions to the problem

$$c \cdot x \to \max, \quad Ax \le b, \tag{1}$$

where $c = (c_1, \dots, c_n)$, $x = (x_1, \dots, x_n)^T$, $A = (a_{ij})$ is an $m \times n$ matrix, and $b = (b_1, \dots, b_m)$.

The set of admissible vectors in this problem (i.e., of $x \in \mathbb{R}^n$ such that $Ax \le b$) is the intersection of finitely many half-spaces. Such set is said to be *polyhedral*. A polyhedral set may be unbounded. A bounded polyhedral set is a *polyhedron*. It is not hard to show that a solution to problem (1) is achieved at a vertex of the polyhedral set. Thus to find the solution one has simply to examine the values of the function to be maximized at all the vertices and to select the largest of them. However, in applied problems the number of vertices is extremely large, and we need a reasonable *procedure of examination*. One such procedure was proposed by Dantzig. We will describe his simplex method in the so-called *nondegenerate* case.

Suppose we have found some vertex $\overline{x}$ (there are effective methods for the search of vertices), and suppose this vertex is nondegenerate. This means that at this vertex exactly n out of m inequalities in the system $A\overline{x} \leq b$ become equalities and the matrix A_n (see below) is nondegenerate. Without loss of generality we may assume that the first n inequalities become equalities; i.e. (writing $a^i = (a_{i1}, \dots, a_{in})$, $i = 1, \dots, m$),

$$a^i \cdot \overline{x} = b_j, \quad 1 \leq i \leq n, \qquad a^i \cdot \overline{x} > b_i, \quad i \geq n+1. \tag{i}$$

By definition, the vertex $\overline{x}$ is nondegenerate if the vectors $\{a^1, ..., a^n\}$ form a basis in $\mathbb{R}^n$, i.e., the matrix $A_n = (a_{ij})_{1 \leq i,j \leq n}$ is nondegenerate. Set $\overline{b} = (b_1, ..., b_n)$. Then we find from the equation $A_n\overline{x} = \overline{b}$ that

$$\overline{x} = A_n^{-1}\overline{b}. \tag{ii}$$

Let $\overline{\lambda}$ be the solution of the equation

$$A_n^T\lambda = c \tag{iii}$$

(where A_n^T is the transpose of A_n). There are two possibilities: $\overline{\lambda} \geq 0$, or some components of $\overline{\lambda}$ are negative.

We will show that in the first case $\overline{x}$ solves problem (1). Indeed, let x be an admissible vector in this problem, i.e.,

$$Ax \leq b. \tag{iv}$$

Set $\overline{\lambda} = (\overline{\lambda}_1, \dots, \overline{\lambda}_n, 0, \dots, 0)$. Then

$$c \cdot x \overset{(iii)}{=} A_n^T\overline{\lambda} \cdot x = \overline{\lambda} \cdot A_n x \overset{(iv)}{\leq} \overline{\lambda} \cdot \overline{b} \overset{(ii)}{=} \overline{\lambda} \cdot A_n\overline{x} = A_n^T\overline{\lambda} \cdot \overline{x} \overset{(iii)}{=} c \cdot \overline{x};$$

hence $\overline{x}$ is a solution to problem (1).

Consider the second possibility. Let, for example, the first coordinate of $\overline{\lambda}$ be negative. Denote by $\overline{y}$ a nontrivial solution of the system

$$a^2 \cdot y = \dots = a^n \cdot y = 0. \tag{v}$$

Since A_n is nondegenerate, we obtain that $a^1 \cdot \overline{y} \neq 0$. We may assume that

$$a^1 \cdot \overline{y} = -\varepsilon < 0. \tag{vi}$$

Then for small $t > 0$ we have

$$a^i \cdot \overline{x} + t\overline{y} < b_i, \quad i = 1,\ i \geq n+1, \qquad a^i \cdot \overline{x} + t\overline{y} = b_i, \quad 2 \leq i \leq n;$$

i.e., the vector $\overline{x} + t\overline{y}$ is admissible in problem (1) and

$$\begin{aligned} c \cdot (\overline{x} + t\overline{y}) &\overset{(v),(vi)}{=} c \cdot \overline{x} + tc \cdot A_n^{-1}(-\varepsilon, 0, \dots, 0)^T \\ &= c \cdot \overline{x} + t(A_n^{-1})^T c \cdot (-\varepsilon, 0, \dots, 0)^T \\ &= c \cdot \overline{x} - t\varepsilon\overline{\lambda}_1 > c \cdot \overline{x}. \end{aligned}$$

Letting t tend to infinity, we find either that the supremum in this problem is infinite or that some jth inequality ($j \geq n+1$) becomes an equality. Then the corresponding point is a vertex and the functional to be maximized takes a larger value at this point than at $\overline{x}$. Taking this vertex for the initial one (provided it is nondegenerate), we repeat the procedure, and so on. This is how the simplex method works in the nondegenerate case.

The simplex method has played an exceptionally important role in the history of numerical methods of optimization. For many years it was unknown whether or not problem (1) is a so-called nonpolynomial (i.e., "difficult") one. In 1970 Klee and Minty constructed examples showing that in some situations the simplex method requires an exponential number of steps. Many mathematicians (including Dantzig himself) used to say that they regarded as a miracle the triumphant "service" of the simplex method in innumerable applied studies. And it was only recently that polynomial algorithms comparable with simplex method in efficiency were constructed.

6. Supplement: Convex analysis in vector spaces

- Convex analysis studies convex sets, convex functions, and convex extremal problems. An important feature of convex objects is that they may be described in different ways, but essentially in geometric and algebro-analytic terms. Moreover, convex objects admit a special calculus, viz., convex calculus. In this section we dispense with finite-dimensional vector spaces and approach the notions considered before from a somewhat different point of view.

6.1. The classes of convex sets.

Here we introduce in geometric terms five classes of convex sets in an arbitrary vector space. The objects of these classes are unions of the so-called elementary objects, which possess certain geometric properties. We assume that the reader is familiar with standard definitions of a subspace, a cone, and a convex set.

Let X be a vector space.

1) By an elementary subspace we mean a straight line passing through the origin. A *subspace* is a union of such lines containing any two lines along with the plane spanned by them. The class of all subspaces of X will be denoted by $\mathrm{Lin}\,(X)$.

2) By an elementary cone we mean a ray going from the origin. A *convex cone* is a union of such rays containing any two rays (unless they go in opposite directions) along with the entire angle between them. The class of all convex cones in X will be denoted by $\mathrm{Cone}\,(X)$.

3) By an elementary zero-convex set we mean a line segment with one end-point the origin. A *zero-convex set* is a union of such segments containing any two segments (unless they go in opposite directions) together with the entire triangle with two sides of these segments. The class of all zero-convex sets in X will be denoted by $\mathrm{Co}_0\,(X)$.

4) By an elementary affine subspace we mean a point. An *affine subspace* is a union of points containing any two different points together with the straight line passing through them. The class of all affine subspaces in X will be denoted by $\mathrm{Aff}\,(X)$.

5) By an elementary convex set we also mean a point. A *convex set* is a union of points containing any two points together with the segment joining them. The class of all convex sets in X will be denoted by $\mathrm{Co}\,(X)$.

By an elementary object we mean any of the elementary sets defined above.

The classes of sets so defined admit a unified algebraic description as classes of sets which along with two points x_1 and x_2 contain all the points of the form $\alpha_1 x_1 + \alpha_2 x_2$, where the pair of real numbers (α_1, α_2) belongs to a certain subset $\mathcal{A} \subset \mathbb{R}^2$: $\mathcal{A} = \mathbb{R}^2$ for $\mathrm{Lin}\,(X)$, $\mathcal{A} = \mathbb{R}^2_+$ for $\mathrm{Cone}\,(X)$, $\mathcal{A} = \{(\alpha_1, \alpha_2) \in \mathbb{R}^2_+ \mid \alpha_1 + \alpha_2 \leq 1\}$ for $\mathrm{Co}_0\,(X)$, $\mathcal{A} = \{(\alpha_1, \alpha_2) \in \mathbb{R}^2 \mid \alpha_1 + \alpha_2 = 1\}$ for $\mathrm{Aff}\,(X)$, and $\mathcal{A} = \{(\alpha_1, \alpha_2) \in \mathbb{R}^2_+ \mid \alpha_1 + \alpha_2 = 1\}$ for $\mathrm{Co}\,(X)$.

6.2. Duality operators.

All the objects in X defined above (subspaces, convex cones, zero-convex sets, affine subspaces, and convex sets), which are unions of elementary objects, have dual objects in the dual space, which are intersections of dual objects to the corresponding elementary ones.

We begin with definitions of dual objects to elementary ones.

Let X' be the space dual to X, i.e., the set of all linear functionals on X. We write $\langle x', x\rangle$ for the action of $x' \in X'$ on $x \in X$.

The dual objects to elementary ones are defined as follows. For the straight line l it is the *annihilator* of l, $l^{\perp} = \{x' \in X' \mid \langle x', x\rangle = 0,\ \forall x \in l\}$ (which is a hyperplane); for a ray r this is the *conjugate cone* of r, $r' = \{x' \in X' \mid \langle x', x\rangle \geq 0,\ \forall x \in r\}$, which is a half-space bounded by a hyperplane passing through the origin; for a segment Δ this is the *polar* of Δ, $\Delta^{\circ} = \{x' \in X' \mid \langle x', x\rangle \leq 1,\ \forall x \in \Delta\}$, which is a translated subspace. We do not specify the dual objects to points as elementary objects, since they will not be used in the sequel.

The bidual elements in X are defined in a natural way: the *biannihilator* of a straight line l, $l^{\perp\perp} = \{x \in X \mid \langle x', x\rangle = 0,\ \forall x' \in l^{\perp}\}$, the *biconjugate cone* of a ray r, $r'' = \{x \in X \mid \langle x', x\rangle \geq 0,\ \forall x' \in r'\}$, the *bipolar* of a segment Δ, $\Delta^{\circ\circ} = \{x \in X \mid \langle x', x\rangle \leq 1,\ \forall x' \in \Delta^{\circ}\}$.

If a is a straight line, a ray, or a segment and d denotes the operation translating a to its annihilator, conjugate cone, or polar respectively, then it is not hard to see that $a^{dd} = a$.

Now we note that the annihilator of the linear hull of two straight lines (or, which is the same, of their sum) is the intersection of the annihilators $((l_1 + l_2)^{\perp} = l_1^{\perp} \cap l_2^{\perp})$. In a similar way, the conjugate cone of the conic hull of two rays (or of their sum) is the intersection of the conjugate cones $((r_1 + r_2)' = r_1' \cap r_2')$. The situation is slightly more complicated for segments. The polar of the convex hull of two segments is indeed the intersection of their polars, but the polar of the sum of segments gives rise to a new operation to be denoted by $\otimes$ $((\Delta_1 + \Delta_2)^{\circ} = \Delta_1^{\circ} \otimes \Delta_2^{\circ} = \bigcup_{0\leq\alpha\leq 1}((1-\alpha)\Delta_1^{\circ} \cap \alpha\Delta_2^{\circ})$. The proofs of all these statements are based on the following simple algebraic facts: if a_1 and a_2 are two real numbers and the expression $\alpha_1 a_1 + \alpha_2 a_2$

— is equal to zero for any α_1, α_2, then $a_1 = a_2 = 0$;

— is nonnegative for any nonnegative α_1, α_2, then $a_1 \geq 0,\ a_2 \geq 0$;

— is no greater than one for any nonnegative α_1, α_2 such that $\alpha_1 + \alpha_2 \leq 1$, then $a_i \leq 1,\ i = 1, 2$;

— is equal to one for any α_1, α_2 such that $\alpha_1 + \alpha_2 = 1$, then $a_1 = a_2 = 1$.

The definition of the new operation on segments is based on one more simple algebraic fact: if a_1 and a_2 are two real numbers and $\alpha_1 a_1 + \alpha_2 a_2 \leq 1$ for $0 \leq \alpha_i \leq 1,\ i = 1, 2$, when $a_1 < 0,\ a_2 \geq 1$, when $a_2 < 0,\ a_1 \geq 1$, and when $a_1 \geq 0,\ a_2 \geq 1,\ a_1 + a_2 \leq 1$, then this sum is $\{a_1 \leq 1\} \otimes \{a_2 \leq 1\}$.

A dual object is defined as the intersection of the sets dual to the elementary ones: if $L \in \mathrm{Lin}\,(X)$, i.e., $L = \bigcup_{l \in L} l$, then $L^\perp = \bigcap_{l \in L} l^\perp = \{x' \in X' \mid \langle x', x\rangle = 0,\ \forall x \in L\}$; if $C \in \mathrm{Cone}(X))$, i.e., $C = \bigcup_{r \in C} r$, then $C' = \bigcap_{r \in C} r' = \{x' \in X' \mid \langle x', x\rangle \geq 0,\ \forall x \in C\}$; if $B \in \mathrm{Co}_0(X)$, i.e., $B = \bigcup_{\Delta \in B} \Delta$, then $B^\circ = \bigcap_{\Delta \in B} \Delta^\circ = \{x' \in X' \mid \langle x', x\rangle \leq 1,\ \forall x \in B\}$.

We see that dual objects belong to the same classes of sets (subspaces, convex cones, and zero-convex sets) as the initial ones but defined in the dual space. Then taking their duals we obtain the objects in X bidual to the initial ones.

6.3. The duality theorem.

The biduality of elementary objects ($a^{dd} = a$) implies that any set is contained in its bidual.

In order to describe the sets which are equal to their bidual sets we will introduce topologies in the vector spaces X and X'. These spaces are *dual* to each other in the sense that the bilinear form $\langle \cdot, \cdot\rangle \colon X' \times X \to \mathbb{R}$, $(x', x) \to \langle x', x\rangle$ has the property that $\langle x', x\rangle = 0$ for any $x \in X$ ($x' \in X'$) implies $x' = 0$ ($x = 0$). This duality enables us to introduce topologies in X' and X making them locally convex topological spaces (LCS). They are called *weak* topologies and are denoted by $\sigma(X, X')$ and $\sigma(X', X)$ respectively. The base of neighborhoods of zero for $\sigma(X, X')$ ($\sigma(X', X)$) consists of finite intersections of the sets $U(x', \varepsilon) = \{x \in X \mid |\langle x', x\rangle| < \varepsilon\}$, $x' \in X'$, $\varepsilon > 0$ ($U(x, \varepsilon) = \{x' \in X' \mid |\langle x', x\rangle| < \varepsilon\}$, $x \in X$, $\varepsilon > 0$). The linear functionals $x \to \langle x', x\rangle$, $x' \in X'$ on X ($x' \to \langle x', x\rangle$, $x \in X$ X') are, obviously, continuous in these topologies, and each of the spaces X and X' is topologically dual to the other one, i.e., $(X, \sigma(X, X'))' = X'$ and $(X', \sigma(X', X))' = X$.

If X is an LCS (in particular, a normed space) and X^* its dual space (i.e., the set of all linear continuous functionals on X), then the bilinear form $(x^*, x) \to \langle x^*, x\rangle$ (where $\langle x^*, x\rangle$ is the value of the functional $x^* \in X^*$ on the element $x \in X$) renders the vector spaces X and X^* mutually (topologically) dual. The topology $\sigma(X, X^*)$ is called weak topology in X (and this is the weakest topology among those in which the functionals in X^* are continuous), and in case X is a normed space, the topology $\sigma(X^*, X)$ is called weak* topology on X^*.

In any LCS the separation theorems for convex sets hold. Here we state the second separation theorem.

THEOREM (second separation theorem). *Let X be an LCS, X^* its dual, A a nonempty convex closed subset of X, and $x_0 \in X \setminus A$. Then the set A and the point x_0 are strictly separable; i.e., there is a nonzero element $x^* \in X^*$ such that* $\sup_{x\in A}\langle x^*, x\rangle < \langle x^*, x_0\rangle$.

The hyperplanes and half-spaces in X and X' are closed in their weak topologies; therefore all dual and bidual sets are also closed. Thus for a set to coincide with its dual it is necessary that it be closed. It can be easily deduced from the second separation theorem that this condition is also sufficient. For example, we will show this for annihilators: *if L is closed, then $L^{\perp\perp} = L$.*

Indeed, as was pointed out, $L \subset L^{\perp\perp}$. Assume that $L \neq L^{\perp\perp}$ and let $x_0 \in L^{\perp\perp} \setminus L$. By the second separation theorem there is a nonzero element $x' \in X'$ such that

$$\sup_{x\in L}\langle x', x\rangle < \langle x', x_0\rangle. \tag{i}$$

Then $x' \in L^{\perp}$ because otherwise $\langle x', \overline{x}\rangle \neq 0$ for some $\overline{x} \in L$ and since $t\overline{x} \in L$ for any $t \in \mathbb{R}$, we arrive at a contradiction with (i). But if $x' \in L^{\perp}$, then the left-hand side of (i) is equal to zero; hence $\langle x', x_0\rangle \neq 0$, which, in its turn, contradicts the fact that $x_0 \in L^{\perp\perp}$.

In fact we have given a two-way description of all convex closed objects. Each of them was first defined *geometrically* (as a set containing two elementary objects together with their linear, conic, or zero-convex combination) and then *algebraically* (as the solution to a system of homogeneous linear equations, homogeneous linear inequalities, or nonhomogeneous inequalities with the same free term). The latter description can be *analytically* expressed as follows: a convex closed set of the given class is equal to its bidual. Thus all these descriptions are contained in the following theorem (which for the finite-dimensional case was proved in Section 2).

THEOREM 1 (on duality of sets). *A subspace L is closed if and only if $L^{\perp\perp} = L$; a convex cone C is closed if and only if $C'' = C$; a zero-convex set B is closed if and only if $B^{\circ\circ} = B$.*

6.4. Convex calculus.

The two basic operations on the class of all convex sets are convex hull of a union ($\operatorname{co}\cup$) (in terms of which subspaces, cones, zero-convex, and simply convex sets are defined) and intersection ($\cap$) (defining the dual objects). Moreover, there is one more important operation, viz., summation ($+$) of two sets. The convex hull of two straight lines coincides with the convex hull of their union and with their sum, so that in $\mathrm{Lin}\,(X)$ there are only two operations: $+$ and its dual, $\cap$. The conic hull of two rays also coincides with the convex hull of their union and with their sum; hence in $\mathrm{Cone}(X)$ there are also only two operations, $+$ and $\cap$.

Further, we should bear in mind that the image and the inverse image of any of our objects under a linear mapping belong to the same class. Moreover, intersections of closed convex sets and inverse images of such sets are closed sets, whereas image, sum, convolution, and convex hull of a union do not in general possess this property. Thus (as will be seen from the forthcoming theorem) the action of the duality operator on the sum, the image, and the convex hull of a union of convex sets yields the desired relations without any additional assumptions, whereas the relations for a similar action on the intersection, the convolution, and the inverse image of convex sets hold under certain conditions, which we express by the sign $\cong$.

THEOREM 2 (formulas of convex calculus of sets).

$$\begin{aligned}
L_i \in \mathrm{Lin}\,(X),\ i=1,2, \implies \quad & (L_1+L_2)^\perp = L_1^\perp \cap L_2^\perp,\\
& (L_1\cap L_2)^\perp \cong L_1^\perp + L_2^\perp,\\
& (AL)^\perp = L^\perp A^*, \quad (LA)^\perp = A^* L^\perp.\\
C_i \in \mathrm{Cone}\,(X),\ i=1,2, \implies \quad & (C_1+C_2)' = C_1' \cap C_2',\\
& (C_1\cap C_2)' \cong C_1' + C_2',\\
& (AC)' = C'A^*, \quad (CA)' \cong A^*C'.\\
B_i \in \mathrm{Co}_0\,(X),\ i=1,2, \implies \quad & (B_1 \operatorname{co}\cup B_2)^\circ = B_1^\circ \cap B_2^\circ,\\
& (B_1\cap B_2)^\circ \cong B_1^\circ \operatorname{co}\cup B_2^\circ,\\
& (AB)^\circ = B^\circ A^*, \quad (BA)^\circ \cong A^*B^\circ,\\
& (B_1+B_2)^\circ \cong B_1^\circ \otimes B_2^\circ,\\
& (B_1\otimes B_2)^\circ \cong B_1^\circ + B_2^\circ.
\end{aligned}$$

There is one more operation, $^+$, which does not appear in this book; hence we do not state the corresponding formulas (although they could be easily written down).

Apart from sets, convex analysis deals with functions. But the theory of convex functions can be easily reduced to that of convex sets because each point $\{x\}$ is associated with a linear function on X, $s_{\{x\}}(x') = \langle x', x\rangle$. The inverse operator is called the subdifferential, $\partial s_{\{x\}}(\cdot) = \{x\}$. The supremum of a family of linear functions is a convex homogeneous (of first order) function (or a sublinear function). The epigraph of a sublinear function is a cone in $X \times \mathbb{R}$. If the function is closed, this cone is closed; hence it is an intersection of half-spaces containing a ray $\{(0, \alpha) \mid \alpha \geq 0\}$. It is not hard to see that in this case this function is a supremum of a family of linear functions, i.e., a support function of some set.

We complete this section by a list of the remaining formulas of convex calculus, which for the finite-dimensional case were given in Section 3. Let X and Y be vector spaces and X' and Y' their dual spaces. All spaces are endowed with the corresponding weak duality topologies.

(1) Let f_1, f_2, g be convex functions on X, f a convex function on Y, $\Lambda\colon X \to Y$ a linear operator, and $\Lambda'\colon Y' \to X'$ the operator conjugate with Λ. Then

1.1 $(f_1 \vee f_2)^* \cong f_1^* \operatorname{co}\wedge f_2^*$; 1.2 $(f_1 \operatorname{co}\wedge f_2)^* = f_1^* \vee f_2^*$;

1.3 $(f_1 \oplus f_2)^* = f_1^* + f_2^*$; 1.4 $(f_1 + f_2)^* \cong f_1^* \oplus f_2^*$;

1.5 $(\Lambda g)^* = g^* \Lambda'$; 1.6 $(f\Lambda)^* \cong \Lambda' f^*$.

(2) Let A_1, A_2, A be convex subsets of X, B a convex subset of Y, and let $\Lambda\colon X \to Y$ be a linear operator. Then

2.1 $\delta(A_1 \cap A_2) = \delta(A_1 \otimes A_2) = \delta A_1 + \delta A_2 = \delta A_1 \vee \delta A_2$;

2.2 $\delta(A_1 \operatorname{co}\cup A_2) = \delta A_1 \operatorname{co}\wedge \delta A_2$; 2.3 $\delta(A_1 + A_2) = \delta A_1 \oplus \delta A_2$;

2.4 $\delta(\Lambda A) = \Lambda\, \delta A$; 2.5 $\delta(B\Lambda) = \delta B\, \Lambda$.

(3) Let A_1, A_2, A, B, Λ be as in the previous item and Λ' the operator conjugate with Λ. Then

3.1 $s(A_1 \cap A_2) \cong sA_1 \oplus sA_2 \cong s\, A_1 \operatorname{co}\wedge sA_2$;

3.2 $s(A_1 \operatorname{co}\cup A_2) = sA_1 \vee sA_2$; 3.3 $s(A_1 + A_2) = sA_1 + sA_2$;

3.4 $s(A_1 \otimes A_2) \cong sA_1 \otimes sA_2$; 3.5 $s(\Lambda A) = sA\, \Lambda'$;

3.6 $s(B\Lambda) \cong \Lambda'\, sB$.

(4) Let A_1, A_2, A, B, Λ be as in the previous item. Then

4.1 $\mu(A_1 \cap A_2) = \mu A_1 \vee \mu A_2$; 4.2 $\mu(A_1 \operatorname{co}\cup A_2) \cong \mu A_1 \oplus \mu A_2$;

4.3 $\mu(A_1 + A_2) \cong \mu A_1 \otimes \mu A_2$; 4.4 $\mu(A_1 \otimes A_2) \cong \mu A_1 + \mu A_2$;

4.5 $\mu(\Lambda A) = \Lambda\, \mu A$; 4.6 $\mu(B\Lambda) = \mu B\, \Lambda$.

CHAPTER 2

Applications

7. Convex analysis of subspaces and cones and the theory of linear equations and inequalities

- Here we show how the theory of linear equations can be developed on the basis of duality and convex calculus of annihilators and the theory of solvability of linear inequalities on the basis of duality and convex calculus of conjugate cones.

Let a system of linear equations $\sum_{i=1}^{n} a_{ij}x_i = b_j$, $j = 1, \dots, m$, or inequalities $\sum_{i=1}^{n} a_{ij}x_i \leq b_j$, $j = 1, \dots, m$, be given, where a_{ij}, $1 \leq i \leq n$, $1 \leq j \leq m$, and b_j, $1 \leq j \leq m$, are given numbers, and x_i, $1 \leq i \leq n$, are unknown variables. The main problems here are: compatibility (solvability) of the systems of equations or inequalities (i.e., the existence of at least one solution), description of the set of all solutions, existence of nonnegative solutions, deriving explicit formulas for solutions, and so on.

Linear equations are one of the most long-standing subjects in mathematics. In the earliest known source containing formulations of mathematical problems, the Rhind papyrus (about 1650 BC), the linear equation $x + \frac{x}{7} = 19$ is discussed. The methods for solving two equations in two unknowns and three equations in three unknowns were described in the Chinese treatise *Mathematics in Nine Books* attributed to the second or first century BC. The general theory of linear equations with finitely many equations and unknowns was built in the 18–19th centuries (G. Cramer, L. Kronecker, and others). In the 20th century many results of the finite-dimensional theory were extended to the infinite-dimensional case (I. Fredholm, F. Riesz). In contrast to the theory of linear equations, the development of the theory of linear inequalities began only in the 19th century.

The set of solutions of a system of linear equations and inequalities is the intersection of half-spaces of the form $\{x \in \mathbb{R}^n \mid a \cdot x \leq b\}$, where $a = (a_1, \dots, a_n)$, $x = (x_1, \dots, x_n)^T$, and $b \in \mathbb{R}$ (in the case of linear equations each hyperplane $\{x \mid a \cdot x = b\}$ may be viewed as the intersection of the half-spaces $\{x \in \mathbb{R}^n \mid a \cdot x \leq b\}$ and $\{x \in \mathbb{R}^n \mid -a \cdot x \leq -b\}$). The intersection of finitely many half-spaces is a *convex polyhedron.* It is clear that a convex polyhedron specified by finitely many homogeneous inequalities, i.e., a set of the form $\{x \in \mathbb{R}^n \mid a^i \cdot x \leq 0,\ i = 1, \dots, m\}$, is a convex cone. Such cones will be called *polyhedral cones.* A cone which is a convex hull of finitely many vectors, i.e., a set of the form $\operatorname{cone}\{b^1, \dots, b^s\}$, is said to be *finitely generated.*

We begin with the famous theorem by H. Weyl (1935) describing the structure of compact convex polyhedra and convex polyhedra that are polyhedral cones.

THEOREM (Weyl).

1. *A convex polyhedron in $\mathbb{R}^n$ is compact if and only if it is the convex hull of a finite number of points.*
2. *A cone in $\mathbb{R}^n$ is polyhedral if and only if it is finitely generated.*

PROOF. 1. Let M_1 be a compact set in $\mathbb{R}^n$ which is the intersection of finitely many half-spaces. Then M_1 is a convex compact set, and by the Minkowski–Krein–Milman theorem (see the Introduction and Section 11) it contains an extreme point. An extreme point of an intersection of half-spaces, being a boundary point, must belong to the intersection of n hyperplanes whose normals form a basis in $\mathbb{R}^n$ (otherwise the intersection of hyperplanes is an affine manifold containing with each point an interval, so the point cannot be extreme). Since there are finitely many collections of n hyperplanes from among the boundaries of our half-spaces, the set M_1 has finitely many extreme points, and by the Minkowski–Krein–Milman theorem it is the convex hull of these points.

Conversely, let the set $M_2 \subset \mathbb{R}^n$ be the convex hull of finitely many points $x^1, \dots, x^N$. Then M_2 is a convex compact set. Without loss of generality we may assume that the origin is an interior point of $M_2 = \operatorname{co}\{x^1, \dots, x^N\}$ (otherwise we restrict our considerations to the affine hull of this set and, if necessary, shift M_2 to contain zero in its interior). Then the polar of M_2 is also a convex compact set (as a closed set contained in some ball), which is the intersection of finitely many half-spaces $\{y \in (\mathbb{R}^n)' \mid y \cdot x^i \leq 1\}$, $i = 1, \dots, N$. By the "only if" statement proved above $M_2^\circ = \operatorname{co}\{y^1, \dots, y^{N'}\}$. By the theorem on bipolar $M_2 = M_2^{\circ\circ}$; hence M_2 is the intersection of finitely many half-spaces $\{x \in \mathbb{R}^n \mid y^i \cdot x \leq 1\}$, $i = 1, \dots, N'$.

2. Let $C = \operatorname{cone}\{a^1, \dots, a^s\} \subset \mathbb{R}^n$. Let us show that C is a polyhedral cone. The case $C = \mathbb{R}^n$ is trivial because, for example, $C = \{x \in \mathbb{R}^n \mid 0 \cdot x \leq 0\}$. Let $C \neq \mathbb{R}^n$. Consider the set $K = \operatorname{co}\{0, a^1, \dots, a^s\}$. By the first part of this theorem there are $c^i \in (\mathbb{R}^n)'$ and $\alpha_i \in \mathbb{R}$, $i = 1, \dots, m$, such that $K = \{x \in \mathbb{R}^n \mid c^i \cdot x \leq \alpha_i,\ 1 \leq i \leq m\}$. Since $0 \in K$, we have $\alpha_i \geq 0$, $1 \leq i \leq m$, and some of these numbers are zeros. Indeed, if $\alpha_i > 0$, $1 \leq i \leq m$, then 0 is an interior point of K and hence of C because, obviously, $K \subset C$. But then C must be the entire $\mathbb{R}^n$, which contradicts the assumption. Let $\alpha_i = 0$, $i = 1, \dots, l$, $l \leq m$. Set $C_1 = \{x \in \mathbb{R}^n \mid c^i \cdot x \leq 0,\ 1 \leq i \leq l\}$. We will show that $C = C_1$. If $x \in C$, then $\alpha x \in K$ for $\alpha > 0$ small enough. Hence $c^i \cdot \alpha x \leq 0$, $i = 1, \dots, l$, so that $x \in C_1$. Conversely, let $x \in C_1$. Then for $\alpha > 0$ small enough the vector αx belongs to K (use the representation of K as an intersection of half-spaces); hence $x \in C$. Thus a finitely generated cone is a polyhedral one.

Now let $C = \{x \in \mathbb{R}^n \mid g^i \cdot x \leq 0,\ i = 1, \dots, m\}$. Consider the cone $C_2 = \operatorname{cone}\{-g^1 \dots, -g^m\}$. It can be easily verified that the cone conjugate with C_2 coincides with C, i.e., $C_2' = C$. As we have proved, the finitely-generated cone C_2 is polyhedral; hence it is closed as an intersection of closed half-spaces. Then by the theorem on bi-conjugate cones $C_2 = C_2'' = C'$. Thus C' is finitely-generated; hence it is a polyhedral cone. By what we have just proved, there is a finitely generated cone C_3 such that $C_3 = C''$. But $C'' = C$, so that C is finitely generated. □

Now we proceed to the solvability problem for the systems of linear equations and inequalities stated above. To this end it well be expedient to write them in a matrix form. Let $A = (a_{ij})$ be an $m \times n$ matrix. The systems of linear equations and inequalities can now be written as $Ax = b$ and $Ax \leq b$, where the inequality between vectors is understood coordinate-wise. The matrix A determines a linear operator from $\mathbb{R}^n$ into $\mathbb{R}^m$, which will be denoted by the same letter. This matrix determines also the linear operator $A' \colon (\mathbb{R}^m)' \to (\mathbb{R}^n)'$ by the rule $A'y = yA$, and it is easily verified that for any $x \in \mathbb{R}^n$ and $y \in (\mathbb{R}^m)'$ the equality $y \cdot Ax = yA \cdot x$ holds.

We denote the image and the kernel of the linear operator A by $\operatorname{Im} A$ and $\operatorname{Ker} A$ respectively. Recall that $L^\perp$ denotes the annihilator of the subspace L.

Convex analysis of subspaces in $\mathbb{R}^n$ consists of the duality theorem: if L is a subspace in $\mathbb{R}^n$, then

(a) $L = L^{\perp\perp}$,

and formulas of convex calculus for annihilators:

(b) $(L_1 + L_2)^\perp = L_1^\perp \cap L_2^\perp$;

(c) $(L_1 \cap L_2)^\perp = L_1^\perp + L_2^\perp$;

if $A\colon \mathbb{R}^n \to \mathbb{R}^m$ is a linear operator, then

(d) $(AL)^\perp = L^\perp A'$;

(e) $(LA)^\perp = A'L^\perp$.

The following result gives the dual description of solvability of linear equations:

THEOREM 1. *Let $A\colon \mathbb{R}^n \to \mathbb{R}^m$ be a linear operator. Then the equation $Ax = b$ is solvable if and only if $b \in (\operatorname{Ker} A')^\perp$.*

PROOF. Solvability of the equation $Ax = b$ means that $b \in A\mathbb{R}^n$. Then

$$A\mathbb{R}^n \overset{(a)}{=} ((A\mathbb{R}^n)^\perp)^\perp \overset{(d)}{=} \left((\mathbb{R}^n)^\perp A'\right)^\perp \overset{def}{=} (\operatorname{Ker} A')^\perp. \qquad \square$$

Convex analysis of polyhedral cones (similar to convex analysis of subspaces) consists of the duality theorem: if C is a polyhedral cone in $\mathbb{R}^n$, then

(a′) $C = C''$,

and formulas of convex calculus of annihilators:

(b′) $(C_1 + C_2)' = C_1' \cap C_2'$;

(c′) $(C_1 \cap C_2)' = C_1' + C_2'$;

if $A\colon \mathbb{R}^n \to \mathbb{R}^m$ is a linear operator, then

(d′) $(AC)' = C'A'$;

(e′) $(CA)' = A'C'$.

With the help of these results we prove three main theorems of the theory of linear inequalities.

THEOREM 2. *Let $A\colon \mathbb{R}^n \to \mathbb{R}^m$ be a linear operator and $b \in \mathbb{R}^m$. Then*

1) *the inequality $Ax \le b$ is solvable if and only if $y \cdot b \ge 0$ for all y such that $y \ge 0,\ A'y = 0$* (Ky Fan (1956));
2) *the equality $Ax = b$ for $x \ge 0$ is solvable if and only if $y \cdot b \ge 0$ for all y such that $A'y \ge 0$* (Minkowski (1896), Farkas (1901));
3) *the inequality $Ax \le b$ for $x \ge 0$ is solvable if and only if $y \cdot b \ge 0$ for all y such that $y \ge 0,\ A'y \ge 0$* (Gale (1960)).

PROOF. We will also use the evident equality $(\mathbb{R}^k_+)' = \mathbb{R}^k_+$.

1) Solvability of the inequality $Ax \le b$ means that $b \in A\mathbb{R}^n + \mathbb{R}^m_+$. The sum of two polyhedral cones in a finite-dimensional space is a polyhedral cone (prove it yourself). Hence the cone $A\mathbb{R}^n + \mathbb{R}^m_+$ is a polyhedral cone in $\mathbb{R}^m$ and then

$$A\mathbb{R}^n + \mathbb{R}^m_+ \overset{(a')}{=} \big((A\mathbb{R}^n + \mathbb{R}^m_+)'\big)' \overset{(b')}{=} \big((\mathbb{R}^n)'A' \cap (\mathbb{R}^m_+)'\big)' = \big(0A' \cap \mathbb{R}^m_+\big)'. \qquad (i)$$

But $0A' = \operatorname{Ker} A'$, so (i) is equivalent to statement 1) of the theorem.

Analogously,

2) The cone $A\mathbb{R}^n_+$ is evidently a polyhedral cone in $\mathbb{R}^m$. Hence

$$A\mathbb{R}^n_+ \overset{(a')}{=} \big((A\mathbb{R}^n_+)'\big)' \overset{(d')}{=} (\mathbb{R}^n_+ A')'. \qquad (ii)$$

We see that (ii) is equivalent to the statement 2) of the theorem.

3) $A\mathbb{R}^n_+ + \mathbb{R}^m_+$ is a polyhedral cone in $\mathbb{R}^m$. Hence

$$A\mathbb{R}^n_+ + \mathbb{R}^m_+ \overset{(a')}{=} \big((A\mathbb{R}^n_+ + \mathbb{R}^m_+)'\big)' \overset{(b'),(d')}{=} (\mathbb{R}^n_+ A' \cap \mathbb{R}^m_+)', \qquad (iii)$$

and (iii) is equivalent to statement 3) of the theorem. □

The following result is an infinite-dimensional generalization of Theorem 1.

Let X be a Banach space. In this case convex analysis of subspaces consists of the duality theorem: L is a closed subspace in X iff $L = L^{\perp\perp}$ (a''), and formulas of convex calculus of annihilators: $(L_1 + L_2)^\perp = L_1^\perp \cap L_2^\perp$ (b''), $(L_1 \cup L_2)^\perp \cong L_1^\perp + L_2^\perp$ (c'') (the equality holds if $L_1^\perp + L_2^\perp$ is a closed subspace). Let A be a linear continuous operator from X to another Banach space Y; then $(AL)^\perp = L^\perp A^*$ (d''), $(LA)^\perp \cong A^* L^\perp$ (e'') (the equality holds if $A^*L^\perp$ is a closed subspace).

THEOREM 1′. a) *Let X and Y be Banach spaces, $A\colon X \to Y$ be a linear continuous operator, $b \in X$, and* $\operatorname{Im} A$ *be a closed subspace in Y. Then the equation $Ax = b$ is solvable if and only if $b \in (\operatorname{Ker} A')^\perp$.*

b) *If $Y = X$ and $A = I - C$, where C is a compact operator, then the following (Fredholm) alternative holds: either*

(i) the equation $Ax = b$ has a unique solution for all $b \in X$, or

(ii) the kernel of A contains a nonzero element.

REMARK. Theorem $1'$ is due to Fredholm and F. Riesz. This was one of the greatest achievements of functional analysis in the 20th century.

PROOF. The statement a) is proved similarly to the corresponding statement of Theorem 1.

For the proof of b) note that if $\operatorname{Ker} A \neq \{0\}$ then nonuniqueness of the solution is obtained already for $b = 0$. Now we will show that if the equation $Ax = b$ is solvable for any $b \in X$, then $\operatorname{Ker} A = \{0\}$. Assume the contrary and let $x_1 \in \operatorname{Ker} A$ be such that $x_1 \neq 0$. By assumption, the equation $Ax = x_1$ is solvable, and if x_2 is its solution, then $x_2 \in \operatorname{Ker} A^2$, because $A^2 x_2 = Ax_1 = 0$. Obviously, $\operatorname{Ker} A \subset \operatorname{Ker} A^2$ strictly since x_2 is in $\operatorname{Ker} A^2$ but not in $\operatorname{Ker} A$, because otherwise we would have $x_1 = 0$. We continue this process to obtain a chain of strict inclusions $\operatorname{Ker} A \subset \operatorname{Ker} A^2 \subset \cdots \subset \operatorname{Ker} A^n \subset \cdots$. Now we use the following fact of convex analysis (which is a trivial corollary of the convex calculus presented in Section 6): *the distance between the unit ball of a normed space and a proper subspace of this space is equal to* 1 (this statement is sometimes referred to as the Riesz theorem on almost perpendicular). This implies that for each $n = 2, 3, \ldots$ there is a vector $e_n \in \operatorname{Ker} A^n$ such that $\|e_n\| = 1$ and $\|x - e_n\| \geq 1/2$ for all $x \in \operatorname{Ker} A^{n-1}$. Consider the sequence $\{Ce_n\}$. Let $m > n$. One can easily check that $z_n = e_n - Ae_n + Ae_m \in \operatorname{Ker} A^{m-1}$. Then $\|Ce_m - Ce_n\| = \|z_n - e_m\| \geq 1/2$; hence it is impossible to select a convergent subsequence from the sequence $\{Ce_n\}$, which contradicts compactness of the operator C. Thus, if the equation $Ax = b$ has a solution for any $b \in X$, then $\operatorname{Ker} A = \{0\}$, and this solution is, obviously, unique. □

8. Classical inequalities, problems of geometry and mechanics

- This section comprises selected problems of analysis and geometry demonstrating the capabilities of convex analysis. Moreover, we consider an extremal problem of mechanics, where convexity does not appear explicitly but in fact plays an important role.

8.1. Some classical inequalities.

There are an innumerable amount of inequalities which are simple consequences of the formulas of convex analysis. The main results of the theory of linear inequalities are also consequences of the theorems of convex analysis. We will illustrate these statements by several examples.

THEOREM (generalized Hölder's inequality). *Let X and Y be locally convex spaces, dual to each other with respect to a bilinear form $\langle \cdot, \cdot \rangle$, and let $p \colon X \to \mathbb{R}$ be a nonnegative continuous sublinear function. Then for any $x \in X$ and $y \in Y$ the following inequality holds:*

$$\langle x, y \rangle \leq p(x)\mu(\partial p)(y). \tag{1}$$

PROOF. Since p is continuous, $\partial p \neq \emptyset$. Let $y \in Y$. By the definition of Minkowski's function, $\mu(\partial p)(y) = \inf\{t > 0 \mid t^{-1}y \in \partial p\}$, so that $\langle x, t^{-1}y \rangle \leq p(x)$ for any $x \in X$ and any $t > 0$, hence formula (1). □

Now we point out some corollaries.

HÖLDER'S INEQUALITY. Let $T = (T, \mu)$ be a space with (nonnegative) σ-finite measure μ, and let $X = L_p(T)$, $1 \leq p < \infty$, $Y = X^* = L_{p'}(T)$ $(1/p + 1/p' = 1)$, and $p(x(\cdot)) = \|x(\cdot)\|_{L_p(T)}$. Since $\partial p = BL_{p'}(T)$ (the unit ball in $L_{p'}(T)$) and $\mu(\partial p)(y(\cdot)) = \|y(\cdot)\|_{L_{p'}(T)}$, we have for any $x(\cdot) \in L_p(T)$ and $y(\cdot) \in L_{p'}(T)$

$$\int_T x(t)y(t)d\mu(t) \leq \|x(\cdot)\|_{L_p(T)} \|y(\cdot)\|_{L_{p'}(T)}.$$

If μ is a discrete measure, we obtain Hölder's inequality for sums (finite or infinite).

INEQUALITIES FOR ARITHMETIC, GEOMETRIC, AND POWER MEANS. Let $x_i > 0$, $i = 1, \dots, n$. The arithmetic and geometric means satisfy the inequality

$$\left(\prod_{i=1}^n x_i\right)^{1/n} \leq \frac{1}{n}\sum_{i=1}^n x_i.$$

This inequality directly follows from the fact that the function $x \to \exp(x)$ is convex and therefore satisfies Jensen's inequality

$$\left(\prod_{i=1}^n x_i\right)^{1/n} = \exp\left(n^{-1}\sum_{i=1}^n \log x_i\right) \leq n^{-1}\sum_{i=1}^n \exp(\log x_i)$$
$$= n^{-1}\sum_{i=1}^n x_i.$$

The inequality for power means

$$\left(\frac{1}{n}\sum_{i=1}^{n} x_i^p\right)^{1/p} \le \left(\frac{1}{n}\sum_{i=1}^{n} x_i^q\right)^{1/q}, \quad 0 < p < q < \infty,$$

follows from Hölder's inequality applied to the space $T = \{1, \dots , n\}$ with counting measure and functions $(x_1^p, \dots , x_n^p)$ and $(1, \dots , 1)$ with q/p in place of p:

$$\sum_{i=1}^{n} x_i^p \le \left(\sum_{i=1}^{n} x_i^q\right)^{p/q} n^{1-p/q} = n\left(n^{-1}\sum_{i=1}^{n} x_i^q\right)^{p/q}.$$

8.2. The classical isoperimetric problem.

THEOREM (on the isoperimetric property of a circle). *Let A be a convex planar figure, $p(A)$ its perimeter, and $S(A)$ its area. Then the following (isoperimetric) inequality holds: $p^2(A) \ge 4\pi S(A)$, which becomes an equality only if $A = \rho B + a$, where $\rho > 0$, B is the unit circle centered at zero, and a is an arbitrary point of the plane.*

There are many ways to prove this result. One of the simplest is as follows. Using Steiner's formula (i), obvious homogeneity relations for p and S, and the Brunn–Minkowski inequality (ii), we obtain,

$$\begin{aligned}&\left(\alpha^2 S(A) + \alpha(1-\alpha)p(A) + \pi(1-\alpha)^2\right)^{1/2}\\ &\qquad\overset{(i)}{=} S^{1/2}\left(\alpha A + (1-\alpha)B\right) \overset{(ii)}{\ge} S^{1/2}(A) + (1-\alpha)\pi^{1/2}.\end{aligned}$$

Squaring this inequality and cancelling similar terms yield the result. □

Now we state its n-dimensional extension.

8.3. Young's inequality.

THEOREM (Young's inequality). *For any compact set $S \subset \mathbb{R}^n$, the diameter $D(S)$ of S and the radius $R(S)$ of the circumscribed sphere are related by the inequality*

$$D(S) \ge \sqrt{\frac{2(n+1)}{n}}\, R(S). \tag{1}$$

Inequality (1) becomes an equality if S is a regular simplex.

PROOF. We have to solve the extremal problem

$$R(S) \to \max, \quad D(S) \leq 1, \tag{1'}$$

where $R(S) = \inf_{x\in\mathbb{R}^n} \max_{y\in S} |x-y|$ and $D(S) = \sup_{x,y\in S} |x-y|$.

The determination of the radius $R(S)$ of the sphere around S consists in finding the infimum of a convex function which is the maximum of convex functions. This fact allows us to apply the clean-up theorem. We state the corresponding assertion as a separate lemma.

LEMMA 1. *For a sphere to be circumscribed around a set S containing more than one point, it is necessary and sufficient that it be circumscribed around some simplex with vertices in S.*

PROOF. Consider the problem

$$f(x) = \max_{y\in S} |x-y| \to \min, \quad x \in \mathbb{R}^n.$$

Clearly f is a continuous convex function growing at infinity; hence there is a vector $\widehat{x}$, where f attains its minimum, i.e., $f(\widehat{x}) = R(S)$. By Fermat's theorem,

$$0 \in \partial f(\widehat{x}). \tag{i}$$

Now we employ the subdifferential form of the clean-up theorem (see Section 3). According to this theorem and (i) there are an integer $1 \leq r \leq n+1$, vectors $z_i \in \mathbb{R}^n$, $y_i \in S$, and positive numbers α_i, $i = 1, \dots, r$, $\sum_{i=1}^r \alpha_i = 1$, such that

$$z_i \in \partial_x|\cdot - y_i|(\widehat{x}) = \partial_x|\widehat{x} - y_i|, \quad |\widehat{x} - y_i| = f(\widehat{x}), \; i = 1, \dots, r,$$

$$\sum_{i=1}^r \alpha_i z_i = 0. \tag{ii}$$

Since S consists of more than one point, $f(\widehat{x}) = |\widehat{x} - y_i| \neq 0$; therefore $\partial_x|\widehat{x} - y_i|$ consists of a single element $(\widehat{x} - y_i)/|\widehat{x} - y_i|$. Together with (ii), this implies that

$$\sum_{i=1}^r \lambda_i \frac{(\widehat{x} - y_i)}{|\widehat{x} - y_i|} = 0 \Longleftrightarrow \widehat{x} = \sum_{i=1}^r \lambda_i y_i. \tag{iii}$$

The vectors $\{y_i\}_{i=1}^r$ are the vertices of a simplex in $\mathbb{R}^n$. Now if we solve the problem on the circumscribed sphere for this simplex, we find that the center of this sphere is the same point $\widehat{x}$ because equation (iii) combined with (ii) provides a sufficient condition for a minimum. □

Thus the proof of Young's inequality reduces to a new variational problem, namely, the problem $(1')$, where S runs over all possible simplexes in $\mathbb{R}^n$. This is not a convex problem any more, but it can be solved directly, without recourse to variational methods (see Ioffe and Tikhomirov (1974)). Its solution is a regular simplex, which completes the proof of Young's inequality. □

8.4. The Fermat–Torricelli–Steiner problem and its extension.

This problem consists in finding for a given simplex in $\mathbb{R}^n$ the point with minimal sum of distances from the vertices of the simplex.

1. Formalization:

$$f(x) = \sum_{k=1}^{n+1} |x - x_i| \to \min, \quad x \in \mathbb{R}^n. \tag{i}$$

This is a convex problem without constraints. Obviously, it has a solution. Denote it by $\widehat{x}$.

2. Lagrange principle. In this case it amounts to

$$0 \in \partial f(\widehat{x}). \tag{ii}$$

3. Analysis. Assuming that $\widehat{x}$ is an interior point of the simplex, relation (ii) can be restated as follows: *the sum of the unit vectors starting from the point $\widehat{x}$ and directed to the vertices of the simplex equals zero.* For $n = 2$ this immediately implies that *the angles between the vectors going from $\widehat{x}$ to the vertices of the triangle are* $120°$.

The point in a triangle from which each side is seen at an angle of $120°$ (the *Torricelli point*) can be constructed with ruler and compass. If such construction is impossible, the required point is the vertex of the obtuse angle (this also immediately follows from Fermat's theorem).

For $n \geq 3$ no constructions of this kind are known. The problem can be solved numerically even for large dimension. However, in the multidimensional case there are versions of this problem which admit explicit solutions. For example: *find the point in $\mathbb{R}^n$ with minimal sum of distances from the origin and the points $e_1 = (1, 0, \ldots, 0), e_2 = (0, 1, \ldots, 0), \ldots, e_n = (0, 0, \ldots, 1)$.*

8.5. The Goddard problem.

The problems considered here are motivated by a problem of rocket dynamics. The motion of a controlled rocket in a homogeneous gravity field is described by the system of equations

$$\dot{x} = v, \quad \dot{v} = \frac{P\omega}{m} + g, \quad \dot{m} = -\frac{P}{k}.$$

Here $x, v \in \mathbb{R}^3$ are the location and velocity vectors of the rocket, ω is a unit vector in $\mathbb{R}^3$, P is the propulsive force, m is the mass of the rocket, g is the vector characterizing the gravitation field (in our case of a homogeneous field, it is a constant vector), and k is a constant describing the dependence of fuel consumption on propulsion.

Set $u = P\omega/m$. The equation $\dot{m} = -P/k$ implies $\omega(\dot{m}/m) = -u/k$. Multiplying both sides of the last equality by ω and using that ω and u are proportional, we obtain

$$\frac{\dot{m}}{m} = -\frac{|u|}{k} \implies m(T_0) = m(0)\exp\left(-\frac{1}{k}\int_0^{T_0} |u|dt\right).$$

Let the location, velocity, and mass of the rocket at the initial time instant $t = 0$ be known. The question is how the rocket should be controlled in order to achieve a given point with a given velocity at a given time T_0 having consumed the minimal amount of fuel. It is easily seen that the formal statement of this problem is as follows:

$$\int_0^{T_0} |u|dt \to \min, \quad \ddot{x} = u + g, \quad x(0) = x_0, \quad \dot{x}(0) = v_0,$$
$$x(T_0) = x_1, \quad \dot{x}(T_0) = v_1.$$

It is immaterial for the solution that x lies in $\mathbb{R}^3$. Henceforth we assume that all the vectors, x, g, and $\dot{x}$, belong to $\mathbb{R}^n$.

We will reduce the problem to the Lyapunov problem in order to use the duality arguments. Since $\ddot{x} = u + g$, we have

$$\dot{x}(t) = v_0 + \int_0^t (u+g)d\tau \implies v_1 = v_0 + \int_0^{t_0} udt + gT,$$
$$x(t) = x_0 + \int_0^t \dot{x}d\tau \implies x_1 = x_0 + v_1T_0 - \frac{gT_0^2}{2} - \int_0^{T_0} tudt.$$

Thus we have obtained the Lyapunov problem

$$\int_0^{T_0} |u|dt \to \min, \qquad \int_0^{T_0} udt = \eta, \quad \int_0^{T_0} tudt = \xi,$$

where

$$\xi = x_0 - x_1 + v_1 T_0 - \frac{gT_0^2}{2}, \quad \eta = v_1 - v_0 - gT_0.$$

Denote by $S(\xi, \eta)$ the value of this problem. It is clear that the function $S(\cdot, \cdot)$ is well defined and finite on the entire $\mathbb{R}^n \times \mathbb{R}^n$. Therefore it is continuous and hence closed. Let us find the conjugate function. According to the definitions,

$$\begin{aligned} S^*(p,q) &= \sup_{(\xi,\eta)} \big(\langle \xi, p\rangle + \langle \eta, q\rangle - S(p,q)\big) \\ &= \sup_{(\xi,\eta)} \left(\langle \xi, p\rangle + \langle \eta, q\rangle - \inf \left(\int_0^{T_0} |u|dt \mid \int_0^{T_0} udt = \eta, \right.\right. \\ &\qquad\qquad\qquad\qquad\qquad\qquad \left.\left. \int_0^{T_0} tudt = \xi \right)\right) \\ &= \sup_{u(\cdot)} \left(\int_0^{T_0} (t\langle u, tp+q\rangle - |u|)dt \right) \\ &= \int_0^{T_0} \sup_u (\langle u, tp+q\rangle - |u|)dt. \end{aligned}$$

It is not hard to see that the last expression equals zero if $|pt+q| \leq 1$ for any $t \in [0, T_0]$ and equals $+\infty$ if there is a point $\tau \in [0, T_0]$ such that $|\tau p + q| > 1$.

Let us describe the set A of the pairs (p, q) satisfying $|tp + q| \leq 1, \ \forall t \in [0, T_0]$. We have

$$tp + q = \frac{t}{T_0}(T_0 p + q) + \left(1 - \frac{t}{T_0}\right) q.$$

Therefore $A = \{(p,q) \mid |T_0 p + q| \leq 1, \ |q| \leq 1\}$.

Now, using the Fenchel–Moreau theorem we can find the value of the problem in an explicit form (we use the notation $r = T_0 p + q \Leftrightarrow$

$p = (r - q)/T_0)$):

$$S(\xi, \eta) = S^{**}(\xi, \eta) = \sup_{(p,q)}(\langle \xi, p\rangle + \langle \eta, q\rangle - S^*(p,q))$$

$$= \sup_{(p,q)\in A} (\langle \xi, p\rangle + \langle \eta, q\rangle) = \sup_{|r|\le 1, |q|\le 1} \left(\left\langle \xi, \frac{r}{T_0}\right\rangle + \left\langle q, \eta - \frac{\xi}{T_0}\right\rangle\right)$$

$$= \left|\frac{\xi}{T_0}\right| + \left|\eta - \frac{\xi}{T_0}\right|$$

$$= \left|\frac{x_1 - x_0}{T_0} - v_1 + \frac{gT_0}{2}\right| + \left|\frac{x_1 - x_0}{T_0} - v_0 - \frac{gT_0}{2}\right|.$$

Obviously, the impulse control

$$\widehat{u}(t) = \left(\frac{x_1 - x_0}{T_0} - v_0 - \frac{gT_0}{2}\right)\delta(t) - \left(\frac{x_1 - x_0}{T_0} - v_1 + \frac{gT_0}{2}\right)\delta(t - T_0),$$

where $\delta(\cdot)$ is Dirac's δ-function is admissible and optimal:

$$\int_0^{T_0} \widehat{u}(t)dt = \eta, \qquad \int_0^{T_0} t\widehat{u}(t)dt = \xi, \qquad \int_0^{T_0} |\widehat{u}(t)|dt = S(\xi, \eta).$$

This control solves the problem. It is not hard to show that in the general case of differently directed vectors ξ and $\eta - \xi/T_0$ the problem has a unique solution. In the degenerate case when ξ and $\eta - \xi/T_0$ are nonzero collinear vectors, there are both continuous and impulse controls (even with a single impulse). The detailed analysis is left to the reader.

9. Kolmogorov-type inequalities for derivatives

- Some parts of the classical calculus and approximation theory are closely related to inequalities for derivatives on the real line and half-line. Here we demonstrate the potentials of convex analysis in finding sharp constants and extremal functions in inequalities of this kind.

By a problem on Kolmogorov-type inequalities for derivatives we mean the one of finding the best possible constant K in a multiplicative inequality of the form

$$\|x^{(k)}(\cdot)\|_{L_q(T)} \le K\|x(\cdot)\|^{\alpha}_{L_p(T)}\|x^{(n)}(\cdot)\|^{\beta}_{L_r(T)} \qquad (K)$$

($0 \le k < n$ are integers, $1 \le p, q, r \le \infty$, $\alpha, \beta \ge 0$, $T = \mathbb{R}$ or $\mathbb{R}_+$), which must hold for all functions $x(\cdot) \in L_p(T)$ with $(n-1)$th derivative locally absolutely continuous on T and $x^{(n)}(\cdot) \in L_r(T)$. We will denote the space of such functions by $\mathcal{W}^n_{p,r}(T)$ or simply by $\mathcal{W}^n_p(T)$ if $p = r$.

For fixed T inequality (K) involves five parameters: n, k, p, q, and r (α and β are uniquely determined by them: $\alpha = (n - k - 1/r + 1/q)/(n - 1/r + 1/p)$, $\beta = 1 - \alpha$). We will denote the sharp (the best) constant in this inequality by $K_T(n, k, p, q, r)$.

In the 60s, S. B. Stechkin found out the important role the exact solutions to such problems play in designing optimal methods for the approximation of unbounded operators by bounded ones and in other problems of the approximation theory. Moreover, such inequalities are useful in the problems of recovery of linear functionals from inaccurate data (see Section 10).

The first sharp constants in inequalities of type (K) were obtained by Landau (1913) ($K_{\mathbb{R}_+}(2, 1, \infty, \infty, \infty) = 2$) and Hadamard (1914) ($K_{\mathbb{R}}(2, 1, \infty, \infty, \infty) = \sqrt{2}$). In the late 30s Kolmogorov (1939) determined the sharp constants in (K) for $T = \mathbb{R}$, $p = q = r = \infty$ *in the general case*, i.e., for any $n \geq 2$, $0 < k < n$. The constants he obtained are $K_{\mathbb{R}}(n, k, \infty, \infty, \infty) = K_{n-k}/K_n^{1-k/n}$, where K_r are the so-called Favard constants (to be discussed below). This result remains one of the most remarkable ones in this area, and the sharp inequalities of type (K) are often referred to as *Kolmogorov's inequalities*.

There have been dozens of papers devoted to the search for the best constants in inequalities of this type for various values of parameters. However, there are only five more cases where complete results comparable to Kolmogorov's were obtained: three of them for $\mathbb{R}$ ($p = q = r = 2$ by Hardy, Littlewood, and Pólya (1934), $p = q = r = 1$ by Stein (1957), and $p = r = 2$, $q = \infty$ by Taikov (1968)); and two for $\mathbb{R}_+$ ($p = q = r = 2$ by Lyubich (1960) and Kuptsov (1975) and $p = r = 2$, $q = \infty$ by Gabushin (1969)).

Apart from the above-mentioned "general" solutions there are dozens of "particular" ones, where the sharp constants are found for specific values of n, k, p, q, and r. Particular solutions have been mostly obtained for low smoothness when $n = 1$ or 2, and a few more for $n = 3$ (see survey papers by Tikhomirov and Magaril-Il'yaev (1985) and Arestov (1996)).

The determination of the sharp constant in inequality (K) is equivalent to finding the exact value of the following extremal problem:

$$\|x^{(k)}(\cdot)\|_{L_q(T)} \to \max, \qquad \|x(\cdot)\|_{L_p(T)} \leq \gamma_1, \quad \|x^{(n)}(\cdot)\|_{L_r(T)} \leq \gamma_2, \tag{1}$$

for a fixed $\gamma_1 > 0$ and $\gamma_2 > 0$.

9.1. On the Lagrange principle.

The problems on inequalities for derivatives are a kind of extremal problem of the calculus of variations and optimal control. In their treatment we will rely on the theory of extremal problems, which will be represented by a very simple and natural device (going back to Fermat and Lagrange).

The reasons behind it are quite elementary. If we are to minimize a function (defined on a normed space) which is differentiable at a point $\widehat{x}$ where it attains a local extremum, then

$$f'(\widehat{x}) = 0. \tag{i}$$

This is Fermat's theorem. We will call condition (i) the *stationarity condition*. In case of a problem with constraints of equality type

$$f_0(x) \to \text{extr}, \qquad f_i(x) = 0, \ 1 \leq i \leq m, \tag{ii}$$

(where f_i are continuously differentiable functions subject to certain regularity conditions), one can seek the extremum using the following approach introduced by Lagrange. In Lagrange's words, slightly adapted, it reads as: *compose the Lagrange function* $\mathcal{L}(x, \lambda) = \sum_{i=0}^{m} \lambda_i f_i(x)$ *and treat it "as if the variables were independent. The equations obtained, together with constraint equations, will serve for the determination of all unknowns"* (the text in the quotation marks is due to Lagrange). The quantities λ_i in the Lagrange function are called the Lagrange multipliers, and $\lambda = (\lambda_1, \ldots, \lambda_m)$ is the set of Lagrange multipliers.

In this connection two comments are in order. First, we can extend the setting to include the problems with inequalities:

$$f_0(x) \to \text{extr}, \qquad f_i(x) \leq 0, \ 1 \leq i \leq m', \quad f_i(x) = 0, \ m'+1 \leq i \leq m. \tag{ii'}$$

For the problems (ii') Lagrange's idea has to be slightly expanded: consider the Lagrange function $\mathcal{L}(x, \lambda) = \sum_{i=0}^{m} \lambda_i f_i(x)$ and, following

Lagrange, treat it as if the variables were independent; i.e., apply Fermat's theorem

$$\mathcal{L}_x(\widehat{x}, \lambda) = 0 \qquad (iii)$$

and supplement the stationarity condition (iii) by nonnegativity conditions on the Lagrange multipliers $\lambda_i \geq 0$, $0 \leq i \leq m'$ (or $\lambda_i \leq 0$ in a maximization problem) and complementary slackness conditions $\lambda_i f_i(\widehat{x}) = 0$ for $m' + 1 \leq i \leq m$.

Secondly, one can add constraints of inclusion type, $x \in A$, to the problem (ii'). Then, assuming that the f_i, $0 \leq i \leq m'$, are convex, the f_i for $i \geq m' + 1$ are affine, and the set A is convex, necessary conditions for $\widehat{x}$ to be the absolute minumum point in this convex problem are nonnegativity and complementary slackness conditions with (iii) replaced by the following minimum principle:

$$\min_{x \in A} \mathcal{L}(x, \lambda) = \mathcal{L}(\widehat{x}, \lambda). \qquad (iii')$$

This theorem was actually proved in Section 5. Let us point out once more an important feature of convex problems: necessary conditions are very close to sufficient. In our case, if $\lambda_0 \neq 0$ and the minimum principle, nonnegativity of the λ_i, and the complementary slackness conditions hold at some admissible point $\widehat{x}$, then $\widehat{x}$ is the absolute minimum point.

But these are only preliminaries to the problems we are about to deal with, viz., the problems of classical calculus of variations and optimal control (including those on inequalities for derivatives). The role of the problem without constraints in the calculus of variations is played by the Bolza problem:

$$\mathcal{B}(x(\cdot)) = \int_{t_0}^{t_1} L(t, x(t), \dot{x}(t))\, dt + l(x(t_0), x(t_1)) \to \text{extr}, \qquad (iv)$$

where $L\colon \mathbb{R} \times \mathbb{R}^n \times \mathbb{R}^n \to \mathbb{R}$, $l\colon \mathbb{R}^{2n} \to \mathbb{R}$ (assuming these functions to be smooth). An analog of Fermat's theorem for the problem (iv) is the following assertion: if $\widehat{x}(\cdot) \in C^1([t_0, t_1], \mathbb{R}^n)$ (a vector function with values in $\mathbb{R}^n$) affords a local minumum in $C^1([t_0, t_1], \mathbb{R}^n)$, then Euler's equation

$$\begin{gathered} -\frac{d}{dt}\widehat{L}_{\dot{x}}(t) + \widehat{L}_x(t) = 0 \\ \big(\text{where } \widehat{L}_{\dot{x}}(t) = L_{\dot{x}}(t, \widehat{x}(t), \dot{\widehat{x}}(t)),\ \widehat{L}_x(t) = L_x(t, \widehat{x}(t), \dot{\widehat{x}}(t))\big) \end{gathered} \qquad (v)$$

holds and the transversality conditions

$$\widehat{L}_{\dot{x}}(t_i) = (-1)^i l_{x(t_i)}(\widehat{x}(t_0), \widehat{x}(t_1)), \quad i = 0, 1, \tag{v'}$$

are fulfilled. (If the integrand L does not depend on t, Euler's equation has the energy integral, $\dot{x}L_{\dot{x}} - L = \text{const.}$)

The problems on inequalities for derivatives for $1 < p, q, r < \infty$ belong to the class of *isoperimetric problems of the calculus of variations*, i.e., the problems of the form

$$\begin{aligned} &\mathcal{J}_0(x(\cdot)) = \int_T |x^{(k)}|^q \, dt \to \max, \\ &\mathcal{J}_1(x(\cdot)) = \int_T |x|^p \, dt \le \delta^p, \quad \mathcal{J}_2(x(\cdot)) = \int_T |x^{(n)}|^r \, dt \le 1. \end{aligned} \tag{vi}$$

In solving problem (vi), one should follow Lagrange's approach. Having rewritten (vi) as

$$\begin{aligned} &\mathcal{J}_0(x_1(\cdot)) \to \max, \quad \mathcal{J}_1(x_1(\cdot)) \le \delta^p, \ \mathcal{J}_2(x_1(\cdot)) \le 1, \\ &\dot{x}_1 = x_2, \ \dot{x}_2 = x_3, \ldots, \dot{x}_n = x_{n+1}, \end{aligned}$$

compose the Lagrange function

$$\begin{aligned} \mathcal{L}(x(\cdot), \lambda) &= \sum_{i=0}^{2} \lambda_i \mathcal{J}_i(x_1(\cdot)) + \int_T \left(\sum_{k=1}^{n} p_k(t)(\dot{x}_k(t) - x_{k+1}(t)) \right) dt \\ &= \int_T L \, dt, \end{aligned}$$

$$L = \lambda_0 |x_k|^q + \lambda_1 |x_1|^p + \lambda_2 |x_{n+1}|^r + \sum_{k=1}^{n} p_k(\dot{x}_k - x_{k+1}),$$

$\lambda = (\lambda_0, \lambda_1, \lambda_2, p_1(\cdot), \ldots, p_n(\cdot))$, $x = (x_1(\cdot), \ldots, x_{n+1}(\cdot))$; write down Euler's equation $-\frac{d}{dt}\widehat{L}_{\dot{x}}(t) + \widehat{L}_x(t) = 0$ and the transversality condition $\widehat{L}_{\dot{x}}(0) = 0$; and select the solution decreasing to zero at infinity.

For $r = \infty$ this is an optimal control problem. It should be rewritten as

$$\begin{aligned} &\mathcal{J}_0(x_1(\cdot)) \to \max \\ &\mathcal{J}_1(x_1(\cdot)) \le \delta^p, \ \dot{x}_1 = x_2, \ \dot{x}_2 = x_3, \ \ldots, \ \dot{x}_n = u, \ |u(t)| \le 1. \end{aligned}$$

Then the Lagrange function will depend on two arguments (x, u) with $x = (x_1(\cdot), \dots, x_n(\cdot))$,

$$\mathcal{L}(x(\cdot), u(\cdot), \lambda) = \sum_{i=0}^{1} \lambda_i \mathcal{J}_i(x_1(\cdot)) + \int_T \left(\sum_{k=1}^{n-1} p_k(t)(\dot{x}_k(t) - x_{k+1}(t)) + p_n(t)(\dot{x}_n(t) - u(t)) \right) dt,$$

$$L = \lambda_0 |x_k|^q + \lambda_1 |x_1|^p + \sum_{k=1}^{n-1} p_k(\dot{x}_k - x_{k+1}) + p_n(\dot{x}_n - u),$$

$\lambda = (\lambda_0, \lambda_1, p_1(\cdot), \dots, p_n(\cdot))$. In this case the condition for a minimum consists of Euler's equation $-\frac{d}{dt}\widehat{L}_{\dot{x}}(t) + \widehat{L}_x(t) = 0$, the transversality condition $\widehat{L}_{\dot{x}}(0) = 0$, convergence to zero at infinity, and the maximum principle

$$\min_{\|u\| \le 1} \mathcal{L}(\widehat{x}(\cdot), u(\cdot), \lambda) = \mathcal{L}(\widehat{x}(\cdot), \widehat{u}(\cdot), \lambda) \iff \min_{|u| \le 1} L(t, \widehat{x}(t), \dot{\widehat{x}}(t), u, \lambda) = \widehat{L}(t).$$

In case $p = \infty$ the Lagrange function instead of $p_1\, dt$ involves a measure $d\mu$ supported on the set where $|x_1(t)|$ achieves the value δ.

We will demonstrate all this by a number of examples.

It is worth noting that the Lagrange principle is applied *heuristically*: we write down the corresponding equations according to this principle without appeal to any exact assertions, then we solve these equations, and finally prove that, indeed, this yields the solution of the problem.

9.2. The Lagrange principle in the problem on inequalities for derivatives.

In this section we consider the problem (1) for $q = \infty$.

1. *Formalization.* Obviously, the problem (1) for $q = \infty$ can be equivalently stated as follows:

$$x^{(k)}(0) \to \max, \quad \|x(\cdot)\|_{L_p(T)} \le \delta, \quad \|x^{(n)}(\cdot)\|_{L_r(T)} \le 1 \tag{2}$$

(where the constants $\gamma_1 = \delta$ and $\gamma_2 = 1$ are again chosen to demonstrate the connection with the recovery problem (see Section 10)). For fixed k and n, (2) is a two-parameter family of problems (depending on p and r); let us associate with each of them the point

$(1/p, 1/r)$ in the square $[0,1]^2$. The problems corresponding to its interior points belong to the classical calculus of variations, the bottom side of the square represents the optimal control problems, on the left-hand side lie the problems of calculus of variations and optimal control with phase constraints, and on the upper and right-hand sides lie the nonsmooth optimization problems. But all of them are convex problems.

2. *The Lagrange principle.* We will write down the Lagrange function for the case of classical calculus of variations ($1 < p, q < \infty$). It has the form

$$\mathcal{L}(x(\cdot), -1, \lambda_1, \lambda_2)$$
$$= -x^{(k)}(0) + \lambda_1\left(\int_T |x(t)|^p dt - \delta^p\right) + \lambda_2\left(\int_T |x^{(n)}(t)|^r dt - 1\right),$$

where we put $\lambda_0 = -1$.

According to the Lagrange principle, if $\widehat{x}(\cdot)$ is the solution to problem (2), there must be $\widehat{\lambda}_1 \geq 0$ and $\widehat{\lambda}_2 \geq 0$ such that the Lagrange function $\mathcal{L}$ attains its minimum at $\widehat{x}(\cdot)$, so that its derivative at this point is equal to zero: $\mathcal{L}_{x(\cdot)} = 0$ (the *stationarity condition*). This leads us to the relation to be referred to as the *basic identity*:

$$x^{(k)}(0) = \widehat{\lambda}_1 p \int_T |\widehat{x}(t)|^{p-1} \operatorname{sign} \widehat{x}(t) x(t) dt + \widehat{\lambda}_2 r$$
$$\times \int_T |\widehat{x}^{(n)}(t)|^{r-1} \operatorname{sign} \widehat{x}^{(n)}(t) x^{(n)}(t) dt,$$

which holds for all admissible $x(\cdot)$. Moreover, the complementary slackness conditions

$$\widehat{\lambda}_1 \left(\|\widehat{x}(\cdot)\|_{L_p(T)} - \delta\right) = 0, \quad \widehat{\lambda}_2 \left(\|\widehat{x}^{(n)}(\cdot)\|_{L_r(T)} - 1\right) = 0$$

must be fulfilled.

If $p = \infty$, then the corresponding term in the Lagrange function is of the form $\lambda_1 \left(\|x(\cdot)\|_{C^b(T)} - \delta\right)$, where $C^b(T)$ is the space of continuous functions on T with the sup-norm. By Fermat's theorem, the fact that the Lagrange function (with $\lambda_i = \widehat{\lambda}_i$, $i = 1, 2$) attains its minimum at $\widehat{x}(\cdot)$ implies that the origin belongs to the subdifferential of this function at this point. Then (by the Moreau–Rockafellar theorem) there is a measure $\widehat{\mu}$ on T such that the basic identity becomes

$$x^{(k)}(0) = \widehat{\lambda}_1 \int_T x(t) d\widehat{\mu}(t) + \widehat{\lambda}_2 r \int_T |\widehat{x}^{(n)}(t)|^{r-1} \operatorname{sign} \widehat{x}^{(n)}(t) x^{(n)}(t) dt.$$

If $r = \infty$, then (2) is an optimal control problem, and we write it as

$$x^{(k)}(0) \to \max, \quad \int_T |x(t)|^p dt \le \delta^p,$$
$$x^{(n)}(\cdot) - u(\cdot) = 0, \quad \|u(\cdot)\|_{L_\infty(T)} \le 1. \tag{2'}$$

The constraints of equality type in this case may be considered as a continuum of numerical equalities; hence the Lagrange function is written as

$$\begin{aligned}\mathcal{L} &= \mathcal{L}((x(\cdot), u(\cdot)), -1, \lambda, p(\cdot)) \\ &= -x^{(k)}(0) + \lambda\left(\int_T |x(t)|^p dt - \delta^p\right) + \int_T p(t)(x^{(n)}(t) - u(t))dt.\end{aligned}$$

If $\widehat{x}(\cdot)$ is a solution to problem (2), then $(\widehat{x}(\cdot), \widehat{u}(\cdot))$ (where $\widehat{u}(\cdot) = \widehat{x}^{(n)}(\cdot)$) is a solution to problem (2′). By the Lagrange principle there are a number $\widehat{\lambda}$ and a function $\widehat{p}(\cdot)$ such that the Lagrange function attains its minimum on the set $\|u(\cdot)\|_{L_\infty(T)} \le 1$ at the point $(\widehat{x}(\cdot), \widehat{u}(\cdot))$. The stationarity condition $\mathcal{L}_{x(\cdot)} = 0$ implies the basic identity

$$x^{(k)}(0) = \int_T \left(\widehat{\lambda} p |\widehat{x}(t)|^{p-1} \operatorname{sign} \widehat{x}(t) x(t) + \widehat{p}(t) x^{(n)}(t)\right) dt,$$

and minimization with respect to $u(\cdot)$ (which in this case reduces to the simple problem $-\int_T \widehat{p}(t)u(t)dt \to \min$, $\|u(\cdot)\|_{L_\infty(T)} \le 1$) yields the equality $\widehat{u}(t) = \operatorname{sign} \widehat{p}(t)$.

Now we will show how the problems can be solved with the aid of these relations. We begin with the simplest example which is a particular case of the problem considered by B. Sz.-Nagy (1941).

EXAMPLE 1. Here $n = 1$, $k = 0$, $p = r = 2$, and let, for definiteness, $T = \mathbb{R}_+$; i.e., we consider the problem

$$x(0) \to \max, \qquad \int_{\mathbb{R}_+} |x(t)|^2 dt \le \delta^2, \quad \int_{\mathbb{R}_+} |\dot{x}(t)|^2 dt \le 1. \tag{3}$$

The basic identity has the form

$$x(0) = 2\lambda_1 \int_{\mathbb{R}_+} \widehat{x}(t)x(t)dt + 2\lambda_2 \int_{\mathbb{R}_+} \dot{\widehat{x}}(t)\dot{x}(t)dt.$$

The complementary slackness conditions are

$$\lambda_1 \left(\|\widehat{x}(\cdot)\|_{L_2(\mathbb{R}_+)} - \delta\right) = 0, \quad \lambda_2 \left(\|\dot{\widehat{x}}(\cdot)\|_{L_2(\mathbb{R}_+)} - 1\right) = 0. \tag{i}$$

Integrating the second term in the basic identity by parts (assuming, of course, that the functions tend to zero at infinity), we arrive at the (Euler) equation, which must be fulfilled for the solution,

$$\lambda_2 \ddot{x} - \lambda_1 x = 0 \tag{ii}$$

with boundary condition

$$2\lambda_2 \dot{x}(0) = -1. \tag{iii}$$

3. *Analysis*. Equation (ii) with conditions (iii) and (i) can be easily solved resulting in the function $\widehat{x}(t) = \sqrt{2\delta}\exp(-t/\delta)$. Putting it into the basic identity we obtain the equality

$$x(0) = \frac{1}{\delta}\int_{\mathbb{R}_+} e^{-t/\delta}x(t)dt - \int_{\mathbb{R}_+} e^{-t/\delta}\dot{x}(t)dt \tag{iv}$$

(which could also be easily verified directly).

We have obtained this relation using the Lagrange principle as a heuristic tool. Now we will show how the solution to problem (3) and the sharp constant in the corresponding Kolmogorov inequality can be derived. Estimating the integrals in the right-hand side of (iv) by the Cauchy–Bunyakovskii inequality, we obtain that any $x(\cdot) \in \mathcal{W}_2^1(\mathbb{R}_+)$ admissible in (3) satisfies the inequality $|x(0)| \le \sqrt{2\delta}$, which is sharp because it becomes an equality for the function $\widehat{x}(\cdot)$. This implies that $\widehat{x}(\cdot)$ is the solution to problem (3). Together with elementary homogeneity arguments, this implies that for all $x(\cdot) \in \mathcal{W}_2^1(\mathbb{R}_+)$ the sharp inequality

$$\|x(\cdot)\|_{C^b(\mathbb{R}_+)} \le \sqrt{2}\|x(\cdot)\|_{L_2(\mathbb{R}_+)}^{1/2}\|\dot{x}(\cdot)\|_{L_2(\mathbb{R}_+)}^{1/2}$$

holds, i.e., $K_{\mathbb{R}_+}(1,0,2,\infty,2) = \sqrt{2}$.

In the next example we solve the general problem on half-lines for $p = r = 2$.

EXAMPLE 2 (Sz.-Nagy (1941)). Here $n = 1$, $k = 0$, and $1 < p, r < \infty$, and we again take $T = \mathbb{R}_+$.

On integrating the basic identity by parts, we obtain Euler's equation

$$\lambda_1 p|x(t)|^{p-1}\operatorname{sign} x(t) - \lambda_2 r\frac{d}{dt}\left(|\dot{x}(t)|^{r-1}\operatorname{sign}\dot{x}(t)\right) = 0, \tag{i}$$

which must be fulfilled for the solution of the corresponding problem, along with the boundary condition $\lambda_2 r|\dot{x}(0)|^{r-1}\operatorname{sign}\dot{x}(0) = -1$. Since the integrands do not depend on time, the energy integral holds:

$$\lambda_1 p|x(t)|^p - \lambda_2(r-1)|\dot{x}(t)|^r = 0. \qquad (ii)$$

(It is established by differentiation, which yields (i). This implies that the left-hand side of (ii) is a constant, and we let it be zero assuming that the functions tend to zero at infinity.) It follows from (ii) that we have to solve the equation $\dot{x} = Ax^{p/r}$. Integrating it we obtain three types of solutions:

$$\widehat{x}(t, C, \alpha) = \begin{cases} C\exp(-\alpha t), & p = r; \\ C(t+\alpha)^{r/(r-p)}, & p > r; \\ C(\alpha - t)_+^{r/(r-p)}, & p < r \end{cases}$$

(where $a_+ = a$ for $a \geq 0$ and 0 for $a \leq 0$). The parameters C and α are determined from the complementary slackness (isoperimetric) conditions $\|\widehat{x}(\cdot)\|_{L_p(\mathbb{R}_+)} = \delta$ and $\|\dot{\widehat{x}}(\cdot)\|_{L_r(\mathbb{R}_+)} = 1$.

For example, in case $p = r$ we obtain that the solution of the corresponding problem is $\widehat{x}(t) = \delta^{1/p'} p^{1/p}\exp(-t/\delta)$ and the basic identity has the form

$$x(0) = \frac{p-1}{\delta}\int_{\mathbb{R}_+}\exp\Big(-\frac{(p-1)t}{\delta}\Big)x(t)\,dt - \int_{\mathbb{R}_+}\exp\Big(-\frac{(p-1)t}{\delta}\Big)\dot{x}(t)\,dt.$$

The subsequent arguments are exactly the same as in the previous example (only with Hölder's inequality used instead of the Cauchy–Bunyakovskii one), and we arrive at the following result.

THEOREM (Sz.-Nagy (1941)). *The following sharp inequality holds:*

$$\|x(\cdot)\|_{C^b(\mathbb{R}_+)} \leq \left(\frac{p+r'}{r'}\right)^{r'/(p+r')} \|x(\cdot)\|^s_{L_p(\mathbb{R}_+)} \|\dot{x}(\cdot)\|^{1-s}_{L_r(\mathbb{R}_+)},$$
$$s = \frac{1}{1+r'/p}.$$

EXAMPLE 3 (Generalized Gabushin's inequality). Gabushin (1969) proved the sharp inequality (K) in case $p = r = 2$, $q = \infty$, and $T = \mathbb{R}_+$ for any $k \in \mathbb{Z}_+$, $n \in \mathbb{N}$, $k < n$. The best constant is expressed in terms of the solution of a system of linear equations.

Here we will consider this setup with k not necessarily an integer. In other words, we will solve the problem

$$\|D^\alpha x(\cdot)\|_{L_\infty(\mathbb{R}_+)} \to \max,$$
$$\|x(\cdot)\|_{L_2(\mathbb{R}_+)} \le \gamma_1, \qquad \|x^{(n)}(\cdot)\|_{L_2(\mathbb{R}_+)} \le \gamma_2,$$

where $D^\alpha x(\cdot)$ denotes the fractional derivative of $x(\cdot)$ of order $\alpha \in \mathbb{R}$. Recall that

$$D^\alpha x(\cdot) := \begin{cases} \dfrac{d^{[\alpha]+1}}{dt^{[\alpha]+1}} \circ I^{[\alpha]+1-\alpha} x(\cdot), & \alpha \ge 0, \\ I^{-\alpha} x(\cdot), & \alpha < 0, \end{cases}$$

where

$$I^\beta x(t) := -\frac{1}{\Gamma(\beta)} \int_t^\infty (\tau - t)^{\beta-1} x(\tau)\, d\tau, \quad \beta > 0,$$

and $\Gamma(\cdot)$ is Euler's gamma function. The derivative D^α is called the *Riemann–Liouville fractional derivative.*

If $\alpha \ge 0$ and the function $x(\cdot)$ is sufficiently "nice" (smooth and rapidly decreasing), then

$$D^\alpha x(t) = -\frac{1}{\Gamma([\alpha] - \alpha + 1)} \int_{\mathbb{R}_+} \tau^{[\alpha]-\alpha} x^{([\alpha]+1)}(\tau + t)\, d\tau, \quad t \ge 0.$$

It is clear that when $\alpha = m \in \mathbb{Z}_+$ this expression reduces to the ordinary derivative $x^{(m)}(t)$, while for $-1 \le \alpha < 0$ it is equal to $I^{-\alpha} x(t)$.

Taking into account this representation it is natural to formalize our initial problem as follows:

$$-\frac{1}{\Gamma([\alpha] - \alpha + 1)} \int_{\mathbb{R}_+} \tau^{[\alpha]-\alpha} x^{([\alpha]+1)}(\tau)\, d\tau \to \max,$$
$$\int_{\mathbb{R}_+} |x(t)|^2 dt \le \gamma_1^2, \quad \int_{\mathbb{R}_+} |x^{(n)}(t)|^2 dt \le \gamma_2^2.$$

This is a convex problem (and at the same time a calculus of variations problem). We proceed as before: write down the Lagrange function and the complementary slackness conditions. Then the stationarity conditions yield the basic identity, which, in turn, implies Euler's equation

$$\lambda_2 x^{(2n)}(t) + \lambda_1 (-1)^n x(t) = (-1)^{n+1} (\Gamma(-\alpha))^{-1} t^{-\alpha-1}$$

and the corresponding boundary conditions. Euler's equation is solved by the method of variation of parameters to yield a unique function satisfying all the relations. Then, as before, we check that this function is indeed the solution of our problem. We suppress these routine derivations and formulate the final result.

THEOREM (generalized Gabushin's inequality). *The inequality*

$$\|D^\alpha x(\cdot)\|_{L_\infty(\mathbb{R}_+)} \le K\|x(\cdot)\|_{L_2(\mathbb{R}_+)}^{(n-\alpha+1/2)/n}\|x^{(n)}(\cdot)\|_{L_2(\mathbb{R}_+)}^{(\alpha-1/2)/n} \qquad (i)$$

holds for all functions $x(\cdot) \in \mathcal{W}_2^n(\mathbb{R}_+)$ *if and only if* $-\frac{1}{2} < \alpha < n - \frac{1}{2}$. *The extremal function in* (*i*) *is*

$$x_0(t) = \frac{1}{\Gamma([\alpha]-\alpha+1)}\left(\sum_{j=1}^{2n} a_j e^{\beta_j t}\int_0^t \tau^{[\alpha]-\alpha}e^{-\beta_j\tau}\,d\tau\right.$$
$$\left. - \sum_{j=n+1}^{2n} a_j e^{\beta_j t}\int_{\mathbb{R}_+}\tau^{[\alpha]-\alpha}e^{-\beta_j\tau}\,d\tau + \sum_{j=1}^{n} b_j e^{\beta_j t}\right),$$

where $\beta_1, \dots, \beta_{2n}$ *are the roots of equation* $z^{2n} + (-1)^n = 0$ ($\operatorname{Re}\beta_j < 0$, $j = 1, \dots, n$), $a_1, \dots, a_{2n}$ *are the unique solutions of the system of equations*

$$\sum_{j=1}^{2n}\beta_j^k\xi_j = \begin{cases} (-1)^{n-[\alpha]}, & k = 2n-[\alpha]-2; \\ 0, & k = 0, 1, \dots, 2n-1,\ k \ne 2n-[\alpha]-2, \end{cases}$$

and $b_1, \dots, b_n$ *are the unique solutions of the system of equations*

$$\sum_{j=1}^{n}\beta_j^k\eta_j = \sum_{j=n+1}^{2n} a_j\beta_j^k\int_{\mathbb{R}_+}\tau^{[\alpha]-\alpha}e^{-\beta_j\tau}\,d\tau, \qquad k = n, \dots, 2n-1.$$

The best constant is

$$K_{\mathbb{R}_+}(n,\alpha,2,\infty,2)$$
$$= \left(\frac{2n}{2(n-\alpha)-1}\right)^{\frac{2(n-\alpha)-1}{4n}}\left(\frac{2n}{2n-1}\right)^{\frac{2(\alpha+1)}{4n}}\|x_0(\cdot)\|_{\mathcal{W}_2^n(\mathbb{R}_+)},$$

where $\|x(\cdot)\|_{\mathcal{W}_2^n(\mathbb{R}_+)} = \|x(\cdot)\|_{L_2(\mathbb{R}_+)} + \|x^{(n)}(\cdot)\|_{L_2(\mathbb{R}_+)}$.

For a more detailed exposition of this result we refer the reader to Magaril-Il'yaev and Tikhomirov (1981).

EXAMPLE 4 (Fuller (1960), Gabushin (1969), Magaril-Il'yaev (1976)). In this example $n = 2$, $k = 0$ or 1, $1 \le p < \infty$, $r = \infty$, and $T = \mathbb{R}_+$. To determine the sharp constant in the corresponding inequality, it suffices to find the solution to problem (2), but there are also other problems equivalent to this one as far as the determination of the sharp constant. To demonstrate various options, consider the following setup:

$$\int_{\mathbb{R}_+} |x(t)|^p dt \to \min, \qquad x^{(k)}(0) = 1, \quad \|\ddot{x}(\cdot)\|_{L_\infty(\mathbb{R}_+)} \le 1, \tag{4}$$

where $k = 0, 1$. This is a convex problem, which is at the same time a problem of optimal control.

First, we consider in detail the case where $k = 0$ and $p = 2$.

1. *Formalization.*

$$\int_{\mathbb{R}_+} x^2(t)dt \to \min,\ x(0) = 1,\ \ddot{x}(\cdot) - u(\cdot) = 0,\ \|u(\cdot)\|_{L_\infty(\mathbb{R}_+)} \le 1. \tag{4'}$$

2. *The Lagrange principle.* The Lagrange function of the problem (4′) is (letting $\lambda_0 = \frac{1}{2}$)

$$\mathcal{L} = \int_{\mathbb{R}_+} \left(\frac{x^2(t)}{2} + p(t)(\ddot{x}(t) - u(t)\right) dt + \mu(x(0) - 1).$$

If $\widehat{x}(\cdot)$ is a solution to problem (4), then $(\widehat{x}(\cdot), \widehat{u}(\cdot))$, where $\widehat{u}(\cdot) = \ddot{\widehat{x}}(\cdot)$, is a solution to problem (4′); and by the Lagrange principle there are $\widehat{\mu}$ and $\widehat{p}(\cdot)$ such that the function $\mathcal{L}$ attains its minimum over the set $\|u(\cdot)\|_{L_\infty(T)} \le 1$ at the point $(\widehat{x}(\cdot), \widehat{u}(\cdot))$. The stationarity condition $\mathcal{L}_{x(\cdot)} = 0$ implies the basic identity

$$\int_{\mathbb{R}_+} (\widehat{x}(t)x(t) + \widehat{p}(t)\ddot{x}(t))\, dt + \widehat{\mu}x(0) = 0, \tag{i}$$

and the minimization of $\mathcal{L}$ with respect to $u(\cdot)$ yields the relation

$$\widehat{u}(\cdot) = \operatorname{sign} \widehat{p}(\cdot). \tag{i'}$$

Integrating the second term of the integrand in (i) by parts and taking into account (i') and the boundary conditions, we obtain

$$\ddot{\widehat{x}}(\cdot) = \operatorname{sign} \widehat{p}(\cdot), \quad \ddot{\widehat{p}}(\cdot) = -\widehat{x}(\cdot), \quad \widehat{p}(0) = 0, \quad \widehat{x}(0) = 1, \tag{ii}$$

and identity (i) is rewritten as

$$\dot{\widehat{p}}(0)x(0) = \int_{\mathbb{R}_+} (\widehat{x}(t)x(t) + \widehat{p}(t)\ddot{x}(t))\,dt \tag{iii}$$

(which could also be verified directly).

Relation (ii) implies that for all $t \in \mathbb{R}_+$ the energy integral

$$\dot{\widehat{p}}(t)\dot{\widehat{x}}(t) + \frac{\widehat{x}^2(t)}{2} = |\widehat{p}(t)| \tag{iv}$$

holds (this is established up to an additive constant by differentiation, the constant being equal to zero because the functions vanish at infinity).

3. *Analysis.* We continue our heuristic line of arguments and set $\dot{\widehat{x}}(0) = -\alpha < 0$ (which undoubtedly is a natural assumption). Now (iv) implies that

$$\dot{\widehat{p}}(0) = \frac{1}{2\alpha} > 0; \tag{v}$$

hence in some neighborhood of zero

$$\widehat{x}(t) = \frac{t^2}{2} - \alpha t + 1. \tag{vi}$$

Because of (ii) and (v) the function $\widehat{p}(\cdot)$ for small t has the form

$$\widehat{p}(t) = -\frac{t^4}{24} + \frac{\alpha t^3}{6} - \frac{t^2}{2} + \frac{t}{2\alpha}. \tag{vii}$$

Let τ be the least positive zero of the function $\widehat{p}(\cdot)$. Substituting it into (vii) we obtain the equation relating α and τ:

$$-\frac{\tau^4}{24} + \frac{\alpha\tau^3}{6} - \frac{\tau^2}{2} + \frac{\tau}{2\alpha} = 0. \tag{viii}$$

To obtain the second relation between these variables we use that the problem is self-similar; i.e., equations (ii) are invariant with respect to transformations $\lambda \to (x_\lambda(\cdot), p_\lambda(\cdot))$, where $x_\lambda(t) = \lambda^2\widehat{x}(t/\lambda)$, $p_\lambda(t) = \lambda^4\widehat{p}(t/\lambda)$. This implies that if $\widehat{x}(\tau) < 0$, then the functions

$$\overline{x}(t) = (\widehat{x}(\tau))^{-1}\,\widehat{x}\left(\sqrt{|\widehat{x}(\tau)|}t + \tau\right),$$
$$\overline{p}(t) = -(\widehat{x}(\tau))^{-2}\,\widehat{p}\left(\sqrt{|\widehat{x}(\tau)|}t + \tau\right)$$

satisfy all the relations in (ii), and self-similarity entails the equalities

$$\widehat{x}(t) = (\widehat{x}(\tau))^{-1}\,\widehat{x}\left(\sqrt{|\widehat{x}(\tau)|}t + \tau\right),$$
$$\widehat{p}(t) = -\,(\widehat{x}(\tau))^{-2}\,\widehat{p}\left(\sqrt{|\widehat{x}(\tau)|}t + \tau\right).$$

Differentiating the first equality at zero and then squaring it, we obtain the relation $\dot{\widehat{x}}^2(\tau) = -\widehat{x}(\tau)\dot{\widehat{x}}^2(0)$, which in view of (vi) is equivalent to the equality $(\tau-\alpha)^2 = -(\frac{1}{2}\tau^2 - \alpha\tau + 1)\alpha^2$. This is a quadratic equation in τ, which yields

$$\tau = \alpha\left(1 + \sqrt{\frac{\alpha^2-2}{\alpha^2+2}}\right). \qquad (ix)$$

It is seen that $\alpha \geq \sqrt{2}$. Substituting the expression for τ into $(viii)$ and denoting the left-hand side of $(viii)$ by $f(\alpha)$, we easily find that $f(\sqrt{2}) = 0$, $f'(\sqrt{2}) = -\infty$, and $f(\alpha) \to +\infty$ as $\alpha \to \infty$. Therefore there exist $\overline{\alpha} > \sqrt{2}$ and $\overline{\tau}$ (related by (ix)), which satisfy the equality $(viii)$.

Now, having these two parameters at our disposal, we construct a solution of system (ii). On the interval $[0, \overline{\tau}]$ we set

$$x_0(t) = \frac{t^2}{2} - \overline{\alpha}t + 1, \quad p_0(t) = -\frac{t^4}{24} + \frac{\overline{\alpha}t^3}{6} - \frac{t^2}{2} + \frac{t}{2\overline{\alpha}}.$$

Self-similarity of the problem enables us to extend these functions to the entire half-line. As a result, we obtain a pair of functions to be denoted by $\overline{x}(\cdot)$ and $\overline{p}(\cdot)$ which satisfy conditions (ii) and are finitely supported, because the joining points of $\overline{x}(\cdot)$ (the zeros of $\overline{p}(\cdot)$) are easily seen to converge to $\overline{\tau}(1-\lambda)^{-1}$, where $\lambda = \sqrt{(\overline{\alpha}^2-2)/(\overline{\alpha}^2+2)}$. Note that $\overline{x}(\cdot)$ is obtained by joining countably many parabolas.[1]

Let us show that the function $\overline{x}(\cdot)$ thus constructed is indeed the solution to problem (4). First, it clearly belongs to $\mathcal{W}^2_{2,\infty}(\mathbb{R}_+)$ and is admissible in problem (4). Next, since $\overline{x}(\cdot)$ and $\overline{p}(\cdot)$ satisfy (ii), one can easily check that identity (iii) (with these functions) holds for all

[1]A process with control making infinitely many switches for a finite time is called a chattering mode. The problem solved by Fuller (1960) was apparently the first example of this kind. By now the theory of chattering control has been thoroughly developed; see, e.g., Zelikin and Borisov (1994).

$x(\cdot) \in \mathcal{W}^2_{2,\infty}(\mathbb{R}_+)$. Then (ii) and (iii) with $x(\cdot) = \overline{x}(\cdot)$ imply that

$$\dot{\overline{p}}(0) - \int_{\mathbb{R}_+} \overline{p}(t)\ddot{x}(t)dt \geq \dot{\overline{p}}(0) - \int_{\mathbb{R}_+} |\overline{p}(t)|dt = \int_{\mathbb{R}_+} \overline{x}^2(t)dt.$$

Now let $x(\cdot)$ be an admissible function in problem (4′). Then using the Cauchy–Bunyakovskii inequality, identity (iii), and the inequality just obtained, we get

$$\begin{aligned}\left(\int_{\mathbb{R}_+} x^2(t)dt\right)^{1/2} &\geq \left(\int_{\mathbb{R}_+} \overline{x}^2(t)dt\right)^{-1/2} \int_{\mathbb{R}_+} \overline{x}(t)x(t)dt \\ &= \left(\int_{\mathbb{R}_+} \overline{x}^2(t)dt\right)^{-1/2} \left(\dot{\overline{p}}(0) - \int_{\mathbb{R}_+} \overline{p}(t)\ddot{x}(t)dt\right) \\ &\geq \left(\int_{\mathbb{R}_+} \overline{x}^2(t)dt\right)^{1/2},\end{aligned}$$

which means that $\overline{x}(\cdot)$ is the solution to problem (4′).

THEOREM (the solution of Fuller's problem for $p = 2$ and $k = 0$). *The following inequality holds:*

$$\|x(\cdot)\|_{C^b(\mathbb{R}_+)} \leq K\|x(\cdot)\|^{4/5}_{L_2(\mathbb{R}_+)}\|\ddot{x}(\cdot)\|^{1/5}_{L_\infty(\mathbb{R}_+)}$$

with the best constant $K_{\mathbb{R}_+}(2, 0, 2, \infty, \infty) = 5^{2/5}2^{-3/5}(3+3\sqrt{33})^{1/10} = 1{,}69659\ldots$.

It is clear that the problem with $k = 1$ can be solved by similar arguments. However, it is not hard to derive its solution from the solution to problem (4′) for $k = 0$ obtained above. Indeed, if $\overline{x}_1(\cdot)$ is a solution to problem (4′) for $k = 1$, then, as before, there is a function $\overline{p}_1(\cdot)$ such that, in particular, the following relations hold:

$$\dddot{\overline{x}}_1(\cdot) = \operatorname{sign} \overline{p}_1(\cdot), \quad \dddot{\overline{p}}_1(\cdot) = -\overline{x}_1(\cdot), \quad \dot{\overline{p}}_1(0) = 0, \quad \dot{\overline{x}}_1(0) = 1. \quad (ii')$$

If $(\overline{x}(\cdot), \overline{p}(\cdot))$ is the pair constructed above when dealing with problem (4) for $k = 0$, then one can easily check that the functions $\overline{x}_1(t) = -\gamma^{-2}\overline{x}(\gamma t + \overline{t})$ and $\overline{p}_1(t) = -\gamma^{-4}\overline{p}(\gamma t + \overline{t})$, where $\overline{t}$ is the unique root of the function $\dot{\overline{p}}(\cdot)$ on the interval $[0, \overline{\tau}]$ and $\gamma = -\dot{\overline{x}}(\overline{t})$ $(\gamma > 0)$, satisfy conditions (ii'); and hence by the same reasons as before, $\overline{x}_1(\cdot)$ is the solution to problem (4) for $k = 1$.

So far we have treated the case $p = 2$. In the general setup when $1 \leq p < \infty$ the solution of the problem can be obtained by exactly the same arguments (except for minor technical complications).

EXAMPLE 5 (Magaril-Il'yaev (1976)). In this example $n = 2$, $k = 0$ or 1, $p = 1$, $1 < r < \infty$, and $T = \mathbb{R}_+$. For determining the sharp constant in the corresponding inequality one can consider, for example, the following problem:

$$\int_{\mathbb{R}_+} \left(|y(t)| + \frac{1}{r}|\ddot{y}(t)|^r\right) dt \to \min, \qquad y^{(k)}(0) = 1, \quad k = 0, 1. \quad (5)$$

This problem turns out to be dual to (4) with respect to the perturbation

$$\int_{\mathbb{R}_+} |x(t)+y(t)|^p dt \to \min, \qquad x^{(k)}(0) = 1, \quad \|\ddot{x}(\cdot)+z(t)\|_{L_\infty(\mathbb{R}_+)} \leq 1,$$

and its solution, up to translations and homothetic transformations, is the function $\overline{p}(\cdot)$ which appeared as a Lagrange multiplier in problem (4). We will not check the duality here, but will simply state the solution to problem (5) based on that of problem (4) already known to us.

We will assume that $k = 0$ and let for simplicity $r = 2$. The Lagrange function of problem (5) has the form ($\lambda_0 = 1$)

$$\mathcal{L} = \int_{\mathbb{R}_+} \left(|y(t)| + \frac{1}{2}\ddot{y}^2(t)\right) dt + \nu(y(0) - 1).$$

If $\widehat{y}(\cdot)$ is the solution to problem (5), then by the Lagrange principle there is $\widehat{\nu}$ such that the Lagrange function attains its minimum at this point; therefore, the stationarity condition $\mathcal{L}_{y(\cdot)} = 0$ implies the identity

$$\int_{\mathbb{R}_+} \left(\operatorname{sign} \widehat{y}(t) y(t) + \ddot{\widehat{y}}(t) y(t)\right) dt + \widehat{\nu} y(0) = 0.$$

Integrating by parts and denoting $\widehat{q}(\cdot) = \ddot{\widehat{y}}(\cdot)$ (in order to stress the duality connections with the previous problem), we finally arrive at the relations

$$\ddot{\widehat{q}}(\cdot) = -\operatorname{sign} \widehat{y}(\cdot), \quad \ddot{\widehat{y}}(\cdot) = \widehat{q}(\cdot), \quad \widehat{q}(0) = 0, \quad \widehat{y}(0) = 1 \qquad (i)$$

(and, moreover, $\dot{\widehat{q}}(0) = -\nu$).

Let $\overline{x}(\cdot)$ and $\overline{p}(\cdot)$ be the functions constructed in the previous example, and let $\beta = \widehat{\alpha} - \sqrt{\widehat{\alpha}^2 - 2}$ and $\mu = \overline{p}(\beta)$ (note that $\overline{x}(\beta) = 0$ and $\mu > 0$). Set $\widehat{y}(t) = \mu^{-1}\overline{p}(\mu^{1/4}t + \beta)$ and $\widehat{q}(t) = -\mu^{-1/2}\overline{x}(\mu^{1/4}t + \beta)$. Now it is easy to verify that the functions $\widehat{y}(\cdot)$ and $\widehat{q}(\cdot)$ satisfy

(i), since $\overline{x}(\cdot)$ and $\overline{p}(\cdot)$ satisfy relations (ii) of the previous example. As before, this implies that $\widehat{q}(\cdot)$ is the solution to problem (5).

The case of an arbitrary $1 < r < \infty$ is treated exactly in the same way. Thus we have the following:

THEOREM (Magaril-Il'yaev (1976)). *Let $1 < r < \infty$. Then the following inequality holds:*

$$\|x^{(k)}(\cdot)\|_{C^b(\mathbb{R}_+)} \leq K\|x(\cdot)\|_{L_1(\mathbb{R}_+)}^{(2-k-1/r)/(3-1/r)}\|\ddot{x}(\cdot)\|_{L_r(\mathbb{R}_+)}^{(k+1)/(3-1/r)}, \qquad k = 0, 1,$$

with the sharp constant

$$K_{\mathbb{R}_+}(2, k, 1, \infty, r) = r^{-\frac{k+1}{3r-1}}\left(\frac{(2-k)r-1}{k+1}\right)^{-\frac{(2-k)r-1}{3r-1}}\left(\frac{(k+1)S_{kr}}{(2-k)r+k}\right)^{-\frac{(2-k)r+k}{3r-1}},$$

where S_{kr} is the value of problem (5).

The relationship between the value of problem (5) and the sharp constant in the corresponding inequality is established by a direct calculation.

EXAMPLE 6 (Landau (1913), Arestov (1972)). In this example $n = 2$, $k = 1$, $1 < r \leq \infty$ (the case $r = \infty$ is due to Landau), and $T = \mathbb{R}_+$. Consider first the case $r < \infty$ and let, for simplicity, $r = 2$. We consider the problem

$$\dot{x}(0) \to \max, \qquad \|x(\cdot)\|_{C^b(\mathbb{R}_+)} \leq \delta, \quad \int_{\mathbb{R}_+} \ddot{x}^2(t)dt \leq 1. \tag{6}$$

According to the aforesaid, if $\widehat{x}(\cdot)$ is the solution to problem (6), then there are $\widehat{\lambda}_i \geq 0$, $i = 1, 2$, and a measure $\widehat{\mu}(\cdot)$ on $\mathbb{R}_+$ such that the identity

$$\dot{x}(0) = \widehat{\lambda}_1 \int_{\mathbb{R}_+} x(t)d\widehat{\mu}(t) + 2\widehat{\lambda}_2 \int_{\mathbb{R}_+} \ddot{\widehat{x}}(t)\ddot{x}(t)dt \tag{i}$$

holds along with complementary slackness conditions

$$\widehat{\lambda}_1\left(\|\widehat{x}(\cdot)\|_{C^b(\mathbb{R}_+)} - \delta\right) = 0, \quad \widehat{\lambda}_2\left(\int_{\mathbb{R}_+} \ddot{\widehat{x}}^2(t)dt - 1\right) = 0. \tag{ii}$$

It is natural to assume that the measure $\widehat{\mu}(\cdot)$ is concentrated on the set of points $t \in \mathbb{R}_+$ such that $|\widehat{x}(t)| = \delta$, and if this measure is not compactly supported, then $\ddot{\widehat{x}}(\cdot)$ cannot belong to $L_2(\mathbb{R}_+)$. Thus we assume that the measure $\widehat{\mu}(\cdot)$ is concentrated within some interval $[0,T]$; therefore, $\widehat{x}(\cdot)$ is a constant outside this interval. Then, integrating by parts in the second integral in (i) and setting

$$\widehat{p}(\cdot) = 2\widehat{\lambda}_2\ddot{\widehat{x}}(\cdot), \tag{iii}$$

we obtain

$$\dot{x}(0) = \widehat{p}(T)\dot{x}(T) - \widehat{p}(0)\dot{x}(0) - \dot{\widehat{p}}(T)x(T) + \dot{\widehat{p}}(0)x(0) \\ + \widehat{\lambda}_1 \int_0^T x(t)d\widehat{\mu}(t) + \int_0^T \ddot{\widehat{p}}(t)x(t)dt.$$

Since $x(\cdot)$ is arbitrary, this implies that $\ddot{\widehat{p}}(\cdot) = 0$, $\widehat{p}(0) = -1$, $\widehat{p}(T) = 0$, and $\widehat{\lambda}_1\widehat{\mu}(t) = -\dot{\widehat{p}}(0)\delta(t) + \dot{\widehat{p}}(T)\delta(t-T)$, where $\delta(\cdot)$ is Dirac's δ-function. The first three relations immediately imply that

$$\widehat{p}(t) = -1 + \frac{t}{T}, \tag{iv}$$

so that $\dot{\widehat{p}}(0) = \dot{\widehat{p}}(T) = 1/T$.

The first condition in (ii) and the fact that the measure $\widehat{\mu}(\cdot)$ is concentrated at the points 0 and T imply that $\widehat{x}(0) = -\delta$, $\widehat{x}(T) = \delta$, and $\dot{\widehat{x}}(T) = 0$. Combined with (iii), (iv), and the second condition in (ii), this implies that

$$\widehat{x}(t) = \delta - 2\delta\left(1 - \frac{t}{T}\right)^3$$

for $0 \le t \le T$ and $\widehat{x}(t) = \delta$ for $t \ge T$, where $T = T(\delta) = 2\sqrt[3]{6}\,\delta^{2/3}$.

Now the identity (i) becomes

$$\dot{x}(0) = \frac{x(T) - x(0)}{T} + \int_0^T \widehat{p}(t)\ddot{x}(t)dt. \tag{i'}$$

The proof that $\widehat{x}(\cdot)$ is indeed the solution to problem (6) proceeds along the same lines as before: it is easily verified that the identity (i') holds for all functions $x(\cdot) \in \mathcal{W}^2_{\infty,2}(\mathbb{R}_+)$. Taking the absolute value of the first term in (i') and estimating the second one by the Cauchy–Bunyakovskii inequality, we obtain $|\dot{x}(0)| \le 3\sqrt[3]{\delta/6}$. This inequality is sharp because it becomes an equality for the function $\widehat{x}(\cdot)$. Therefore $\widehat{x}(\cdot)$ is the solution to problem (6).

If $1 < r < \infty$, then we obtain by the same arguments that $\widehat{x}(t) = \delta - 2\delta\left(1 - \frac{t}{T}\right)^{r'+1}$, $0 \le t \le T$, where T depends on r and δ. If $r = \infty$, i.e., $r' = 1$, the expression for the extremal function remains valid in this case as well. Thus the following theorem holds.

THEOREM (Landau–Arestov). *Let* $1 < r \le \infty$. *Then*

$$\|\dot{x}(\cdot)\|_{C^b(\mathbb{R}_+)} \le K \|x(\cdot)\|_{C^b(\mathbb{R}_+)}^{1/(r'+1)} \|\ddot{x}(\cdot)\|_{L_r(\mathbb{R}_+)}^{r'/(r'+1)},$$

where

$$K_{\mathbb{R}_+}(2,1,\infty,\infty,r) \\ = 2^{1/(r'+1)}((r'+1)/r')^{r'/(r'+1)} \ (K_{\mathbb{R}_+}(2,1,\infty,\infty,\infty) = 2).$$

EXAMPLE 7 (Magaril-Il'yaev (1983)). Here $n = 2$, $k = 0$, $1 \le p < \infty$, and $r = 1$; i.e., we are to solve the following problem (assuming for definiteness that $T = \mathbb{R}_+$):

$$y(0) \to \max, \qquad \int_{\mathbb{R}_+} |y(t)|^p dt \le \delta^p, \quad \int_{\mathbb{R}_+} |\ddot{y}(t)| dt \le 1. \tag{7}$$

This problem is equivalent to the one dual to (6) in the same sense as problem (5) is dual to (4). However, problem (7) has no solution. But it can be extended by replacing $L_1(\mathbb{R}_+)$ by the space of measures on $\mathbb{R}_+$, and then the resulting problem has a solution, which can be obtained using the solution of the previous one. Without presenting this derivation we will only state the final result.

THEOREM (Magaril-Il'yaev (1983)). *Let* $1 \le p < \infty$. *Then*

$$\|x(\cdot)\|_{C^b(T)} \le K \|x(\cdot)\|_{L_p(T)}^{1/(1+1/p)} \|\ddot{x}(\cdot)\|_{L_1(T)}^{(1/p)/(1+1/p)},$$

where $K_{\mathbb{R}}(2,0,p,\infty,1) = ((p+1)/8)^{1/(p+1)}$ *and* $K_{\mathbb{R}_+}(2,0,p,\infty,1) = (p+1)^{1/(p+1)}$.

10. Convex analysis and extremal problems of approximation and recovery

- Recovery is one of the central topics of numerical analysis. Its meaning can be briefly explained as follows: we are interested in certain data about an object and we have to recover them (with the best possible accuracy) from the available information (typically inaccurate and/or incomplete). Specifically, we will primarily be interested in *optimal recovery of linear functionals on spaces of polynomials and classes of smooth and analytic functions from some information on these functions.*

Under certain conditions the error in optimal recovery of a linear functional is equal to the value of a certain convex extremal problem, specifically, a maximization problem of a linear functional, and hence it equals the value (on this functional) of the support function of a certain set. Thus from the point of view of the general theory this subject is just an illustration of the use of the calculus formulas for support functions. But in fact we will not use this calculus. The maximization problem of a linear functional is in some cases dual to the "distance problem", which is a classical problem of the approximation theory, and in other cases to the "problem on inequalities" (for polynomials or functions from some smoothness classes). If the values of these problems are equal on some admissible elements, then these elements are solutions to these problems (which will be the central point of our arguments). Moreover, this entails the description of the optimal method in the corresponding recovery problem.

10.1. Introduction.

Before stating the optimal recovery problem in a general form, we will explain it by a simple example of the interpolation problem. Let t_i, $i = 1, \dots, n$, be some fixed points in an interval $[a, b]$. Suppose we know the values $x(t_1), \dots, x(t_n)$ of a continuous function $x(\cdot)$ on $[a, b]$ and we are interested to know the value of this function at an intermediate point $\tau \in [a, b]$. Obviously, the problem makes no sense in the class of all functions: $x(\tau)$ can take any value for given $x(t_i)$, $i = 1, \dots, n$. But usually the researcher has some information of interest about the function $x(\cdot)$. Generally this information consists of the knowledge that $x(\cdot)$ belongs to a certain set (class) C of continuous functions on $[a, b]$. The optimal recovery problem of the value of $x(\cdot)$ at the point τ over the class C is stated as follows. Let $x(\cdot) \in C$ and let φ be an arbitrary function on $\mathbb{R}^n$. We consider the value $\varphi(x(t_1), \dots, x(t_n))$ as an estimate for $x(\tau)$ and characterize the accuracy of this estimation method by the quantity

$$e(x(\tau), C, F_{\{t_1,\dots,t_n\}}, \varphi) = \sup_{x(\cdot) \in C} |x(\tau) - \varphi(x(t_1), \dots, x(t_n))|, \quad (*)$$

where $F_{\{t_1,\dots,t_n\}}$ is the operator which associates with $x(\cdot)$ the set of values $\{x(t_1), x(t_2), \dots, x(t_n)\}$. In other words, $(*)$ is the largest error which we can get on the class C using this estimation method. Now we minimize the error $(*)$ over all functions (methods) φ; i.e., we consider the quantity

$$E(x(\tau), C, F_{\{t_1,\dots,t_n\}}) = \inf_{\varphi} e(x(\tau), C, F_{\{t_1,\dots,t_n\}}, \varphi). \qquad (**)$$

The method $\widehat{\varphi}$ which affords the minimum value in $(**)$ is called the *optimal method of recovery of* $x(\tau)$ *on the class* C.

It may happen (say, due to measurement errors) that the values $x(t_1), \dots, x(t_n)$ are not known precisely and we have numbers $y_1, \dots, y_n$, which we know, say, to differ from the true values by at most $\delta > 0$, i.e., $|y_i - x(t_i)| \le \delta$, $i = 1, \dots, n$. In this case we take some function $\varphi(y_1, \dots, y_n)$ as an estimate for $x(\tau)$ and (similarly to $(*)$) characterize the accuracy of this method by the quantity

$$\begin{aligned} &e_\delta(x(\tau), C, F_{\{t_1,\dots,t_n\}}, \varphi) \\ &\quad = \sup_{\substack{x(\cdot)\in C,\ (y_1,\dots,y_n)\in\mathbb{R}^n \\ |y_i - x(t_i)|\le\delta}} |x(\tau) - \varphi(y_1, \dots, y_n)|. \end{aligned} \qquad (***)$$

Then we again minimize over all φ and call the minimizer the *optimal method of recovery.*

The set over which the supremum in $(***)$ is taken may be written in a different form. Denote by F the multi-valued mapping from C to $\mathbb{R}^n$ which associates with each $x(\cdot) \in C$ the set $Fx(\cdot) = \{y = (y_1, \dots, y_n) \mid |y_i - x(t_i)| \le \delta,\ i = 1, \dots, n\}$. Then we can say that the supremum in $(***)$ is taken over $x(\cdot) \in C$ and $y \in \mathbb{R}^n$ such that $y \in Fx(\cdot)$.

Now we summarize the aforesaid in somewhat different terms. We are given a set (class) C in the vector space X of all continuous functions on the interval $[a, b]$. Our information about $x(\cdot) \in C$ is given by $Fx(\cdot)$, where F is a mapping from C to $\mathbb{R}^n$ which associates with $x(\cdot)$ either the vector $(x(t_1), \dots, x(t_n))$ (in which case the information is said to be precise, or error-free) or a set; i.e., F is a multi-valued mapping (in which case the information is inaccurate, subject to an error). On X a linear functional x' is given, namely, $\langle x', x(\cdot)\rangle = x(\tau)$. The optimal recovery problem of the functional x' on the class C based on information F is formulated as follows. Each function φ on $\mathbb{R}^n$ is said to be a recovery method (of x' in C based on information F), and the accuracy of this method of estimation is measured by the quantity $(***)$, which for $\delta = 0$, i.e., in case of precise information, coincides with $(*)$. We are interested in the minimal value of this quantity over all methods, which is called the optimal recovery error, as well as in the method $\widehat{\varphi}$ which affords this minimum, called the method of optimal recovery (of x' in C based on information F).

It is clear that in this problem $x(\tau)$ may be replaced by the integral of $x(\cdot)$ over $[a, b]$, by the value of a derivative of $x(\cdot)$ at a point τ (provided the functions in C are sufficiently smooth), and so on. Moreover, instead of the information on the values of the function at given points, one may have, e.g., the (precise or approximate) values of some moments of $x(\cdot)$ or of its Fourier coefficients (if it is periodic), and so on.

The above discussion leads us in a natural way to the following general setting of the problem. Let C be a set (class) in a real or complex vector space X. About each element $x \in C$ we have some information $F(x)$, where F is the *information mapping* from C to another real or complex vector space Y. The information on elements in C may be inaccurate; hence F is, in general, a multi-valued mapping. Next, let x' be a linear functional on X. The problem is to recover (in the best possible way) the functional x' on the class C based on the available information F. This is formalized as follows: any function $\varphi\colon Y \to \mathbb{R}(\mathbb{C})$ is considered as a method of recovery (of the functional x' on the class C based on information F). The *error* of this method is

$$e(x', C, F, \varphi) = \sup_{x \in C,\, y \in F(x)} |\langle x', x\rangle - \varphi(y)|.$$

The value of the problem

$$e(x', C, F, \varphi) \to \min \tag{1}$$

(to be denoted by $E(x', C, F)$), where the minumum if taken over all possible methods of recovery is called the *optimal recovery error*, and any solution $\widehat{\varphi}$ of this problem is called an *optimal recovery method.* We will write it as $\langle x', x\rangle \approx \widehat{\varphi}(y)$, $y \in F(x)$.

The problem (1) will be called the *optimal recovery problem.* In the case where F is a linear operator and Y is finite-dimensional this problem was considered by Smolyak (1965). He proved that if C is convex and centrally symmetric,[2] then among optimal methods there is a linear one. This statement remains valid for arbitrary vector spaces X and Y and multi-valued mapping F such that its graph $\operatorname{gr} F = \{(x, y) \in C \times Y \mid y \in F(x)\}$ is a convex centrally symmetric set (Magaril-Il'yaev and Osipenko (1991)). In this case $E(x', C, F)$ is

[2] A set C in a vector space is called centrally symmetric if $C = -C$.

the solution to the convex problem

$$\sup_{x\in A,\, y\in F(x)} |\langle x', x\rangle - \langle y', y\rangle| \to \min, \quad y \in Y'. \tag{1'}$$

This problem (assuming henceforth for definiteness that all the vector spaces are real) is dual to the following extremal problem:

$$\langle x', x\rangle \to \max, \qquad x \in F^{-1}(0), \quad x \in C \tag{2}$$

(where $F^{-1}(y) = \{x \in X \mid y \in F(x)\}$ is the inverse image of y under the mapping F; if F is single-valued, then the inclusion $x \in F^{-1}(0)$ is equivalent to $F(x) = 0$) with respect to the perturbation (see Section 5)

$$\langle x', x\rangle \to \max, \qquad x \in F^{-1}(y), \quad x \in C.$$

We call problem (2) the *associated problem* (with the optimal recovery problem (1)) and denote its value by $S(x', A, F)$.

Note the following general fact: *if C and the graph* $\operatorname{gr} F = \{(x,y) \in X \times Y \mid y \in F(x)\}$ *of the mapping F are centrally symmetric sets and the set of admissible elements in problem* (2) *is nonempty, then*

$$E(x', C, F) \geq S(x', C, F). \tag{3}$$

Indeed, let x be an admissible element of (2); then $-x$ is also an admissible element and we have for any $\varphi\colon F(C) \to \mathbb{R}$

$$\begin{aligned} 2\langle x', x\rangle &\leq |\langle x', x\rangle - \varphi(0) + \varphi(0) + \langle x', x\rangle| \\ &\leq |\langle x', x\rangle - \varphi(0)| + |\langle x', -x\rangle - \varphi(0)| \\ &\leq 2 \sup_{x\in F^{-1}(0)} |\langle x', x\rangle - \varphi(0)| \\ &\leq 2 \sup_{x\in C,\, y\in F(x)} |\langle x', x\rangle - \varphi(y)|, \end{aligned}$$

which easily implies (3).

Note that many extremal problems of the approximation theory or dual to them have the form (2), so that each of them is uniquely associated with some optimal recovery problem. The simultaneous treatment of these problems leads to new results and sometimes to interesting interpretations of well-known facts. After the pioneering paper by Smolyak (1965) mentioned above the subject of optimal recovery of linear functionals was further explored in various directions; see Osipenko (1976), (2000), Micchelli and Rivlin (1977), (1985),

Arestov (1989), Plaskota (1996), Magaril-Il'yaev and Tikhomirov (1997), and the references therein.

10.2. General results.

The treatment of specific problems will rely on two simple general results to be given in the two subsequent subsections.

I. *The problem on the distance from a point to a subspace and recovery based on precise information.* We begin with the problem of finding the distance from a point to a subspace in a normed space because it is the earliest problem in the approximation theory. Historically the first problem of this kind was the problem of approximating a continuous function by polynomials. It was posed and solved for a number of special cases by P. L. Chebyshev.

Let X be a normed space and L a subspace in X. Consider the problem of approximation of an element $x_0 \in X$ by the subspace L:

$$\|x_0 - \xi\|_X \to \min, \quad \xi \in L. \tag{P_I}$$

Denote by $d(x_0, L, X)$ the value of this problem, which is the distance between x_0 and L. The solution $\widehat{\xi}$ of problem (P_I) will be referred to as the *element of best approximation* (of x_0 by the subspace L). Problem (P_I) is convex. The dual problem with respect to the perturbation

$$\|x_0 + x - \xi\|_X \to \min, \quad \xi \in L$$

(see Section 5) has the form

$$\langle x^*, x_0\rangle \to \max, \qquad x^* \in L^\perp, \quad x^* \in BX^*. \tag{P_I^*}$$

Recall that BX^* is the unit ball in X^* and $L^\perp$ the annihilator of L. (Thus problem (P_I^*) consists in the determination of $s(BX^* \cap L^\perp)$, the support function of the intersection of BX^* and $L^\perp$.)

Problem (P_I^*) may be viewed as the one associated with the optimal recovery problem of the linear functional $x^* \to \langle x^*, x_0\rangle$ on the class BX^* based on the following information: for each $x^* \in BX^*$ we know its restriction to L; i.e., the information operator is $F_L\colon BX^* \to L^*$, $F_L x^* = x^*|_L$.

Primary attention will be directed to the case of finite-dimensional L (in which case we write L_n for L, where $n = \dim L$). Let $x_1, \dots, x_n$ be a basis in L_n. Then problem (P_I^*) is equivalent to

$$\langle x^*, x_0\rangle \to \max, \qquad \langle x^*, x_i\rangle = 0, \quad i = 1, \dots, n, \quad x^* \in BX^*. \tag{$P_{I_n}^*$}$$

This is the associated problem to the recovery problem of the linear functional $x^* \to \langle x^*, x_0 \rangle$ on the class BX^* with information operator $F_{L_n} : BX^* \to \mathbb{R}^n$, $F_{L_n}x^* = (\langle x^*, x_1 \rangle, \dots, \langle x^*, x_n \rangle)$.

A similar problem on the distance may be considered in the dual space X^* of a normed space X:

$$\|x_0^* - \eta\|_{X^*} \to \min, \quad \eta \in M, \qquad (P_I')$$

where $x_0^* \in X^*$ and M is a subspace of X^*.

The dual problem to (P_I') (with respect to a perturbation similar to the above) has the form

$$\langle x_0^*, x \rangle \to \max, \qquad x \in M^\perp, \quad x \in BX, \qquad (P'^*_I)$$

where $M^\perp = \{x \in X \mid \langle x^*, x \rangle = 0, \ \forall x^* \in M\}$.

Problem (P'^*_I) is the associated problem to the recovery problem of the linear functional x_0^* on the class BX based on the following information: we know about each element $x \in BX$ that all the functionals in M vanish on this element.

If $\dim M = n$ (in which case we write M_n for M) and $x_1^*, \dots, x_n^*$ is a basis in M, then problem (P'^*_I) is equivalent to

$$\langle x_0^*, x \rangle \to \max, \qquad \langle x_i^*, x \rangle = 0, \ i = 1, \dots, n, \quad x \in BX. \quad (P'^*_{I_n})$$

This problem is associated with the recovery problem of the functional x_0^* on the class BX with information operator $F_{M_n} : BX \to \mathbb{R}^n$, $F_{M_n}x = (\langle x_1^*, x \rangle, \dots, \langle x_n^*, x \rangle)$.

In what follows, when dealing with specific problems, we will repeatedly use the following results.

THEOREM I. *If the elements $\widehat{\xi}$ $(\widehat{\eta})$ and $\widehat{x}^*$ $(\widehat{x})$ are admissible in problems (P_I) $((P_I'))$ and (P_I^*) $((P'^*_I))$ respectively and on these elements the equality between the values of these problems is attained,*

$$\langle \widehat{x}^*, x_0 \rangle = \|x_0 - \widehat{\xi}\|_X, \quad (\langle x_0^*, \widehat{x} \rangle = \|x_0^* - \widehat{\eta}\|_{X^*}), \qquad (I)$$

then $\widehat{\xi}$ $(\widehat{\eta})$ is a solution to problem (P_I) $((P_I'))$, and $\widehat{x}^*$ $(\widehat{x})$ is a solution to problem (P_I^*) $((P'^*_I))$. Moreover, the linear functional $x^* \to \langle x^*, \widehat{\xi} \rangle$ $(x \to \langle \widehat{\eta}, x \rangle)$ is the optimal method in the corresponding recovery problem, i.e, $\langle x^*, x_0 \rangle \approx \langle x^*, \widehat{\xi} \rangle$ $(\langle x_0^*, x \rangle \approx \langle \widehat{\eta}, x \rangle)$ and $E(x_0, BX^*, F_L) = \langle \widehat{x}^*, x_0 \rangle$ $(E(x_0^*, BX, F_M) = \langle x_0^*, \widehat{x} \rangle)$.

PROOF. We will give the proof for the problem on the distance in the initial space X; for the dual space the arguments are similar. Let ξ and x^* be arbitrary admissible elements in problems (P_I) and (P_I^*) respectively. Using consecutively the facts that $x^* \in L^\perp$, $\|x^*\|_{X^*} \le 1$, equality (I), $\widehat{x}^* \in L^\perp$, and $\|\widehat{x}^*\|_{X^*} \le 1$, we obtain

$$\begin{aligned}\langle x^*, x_0\rangle &= \langle x^*, x_0 - \widehat{\xi}\rangle \le \|x_0 - \widehat{\xi}\|_X \\ &= \langle \widehat{x}^*, x_0\rangle = \langle \widehat{x}^*, x_0 - \xi\rangle \le \|x_0 - \xi\|_X.\end{aligned}$$

This implies that $\langle x^*, x_0\rangle \le \langle \widehat{x}^*, x_0\rangle$ for any element x^* admissible in (P_I^*) and $\|x_0 - \widehat{\xi}\|_X \le \|x_0 - \xi\|_X$ for any element ξ admissible in (P_I). Thus $\widehat{x}^*$ is a solution to problem (P_I^*) and $\widehat{\xi}$ is a solution to problem (P_I).

Next, using (3), the equality between the values of the problems already obtained, and some elementary inequalities, we get

$$\begin{aligned}E(x_0, BX^*, F_L) &\ge S(x_0, BX^*, F_L) = d(x_0, L, X) = \|x_0 - \widehat{\xi}\|_X \\ &\ge \sup_{x^* \in BX^*} |\langle x^*, x_0\rangle - \langle x^*, \widehat{\xi}\rangle| \ge E(x_0, BX^*, F_L);\end{aligned}$$

i.e., all these inequalities are in fact the equalities; hence the functional $x^* \to \langle x^*, \widehat{\xi}\rangle$ is the optimal recovery method. □

II. *The norms of linear functionals on subspaces, inequalities for derivatives, and recovery based on inaccurate information.* The problems considered here originate from Chebyshev's problem on extrapolation of algebraic polynomials (*from among algebraic polynomials of degree at most n and not exceeding $\delta > 0$ in absolute value on the interval $[-1,1]$, find the one with maximal value at a point τ $(|\tau| > 1)$*), as well as from problems on inequalities for derivatives of algebraic and trigonometric polynomials.

Let X and Y be normed spaces, X^* the space dual to X, $x^* \in X^*$, $\Lambda\colon X \to Y$ a linear continuous operator, and L a subspace of X. Consider the problem

$$\langle x^*, x\rangle \to \max, \qquad \|\Lambda x\|_Y \le \delta, \quad x \in L. \qquad (P_{II})$$

(This problem, similarly to (P_I^*), consists in finding the support function of the intersection of L with the inverse image under the mapping Λ of the ball of radius δ centered at zero.) This problem is associated with the recovery problem of a linear functional x^* on subspace L based on information $F\colon X \to Y$, $Fx = \Lambda x + \delta BY$, where BY is the unit ball in Y. In other words, for each $x \in L$ we know an element $y \in Y$ distant from Λx by at most δ in the metric of Y.

The dual problem to (P_{II}) with respect to the perturbation

$$\langle x^*, x\rangle \to \max, \qquad \|\Lambda x + y\|_Y \le \delta, \quad x \in L,$$

has the form

$$\delta\|y^*\|_{Y^*} \to \min, \quad \Lambda^* y^*|_L = x^*|_L, \tag{P_{II}^*}$$

where Y^* is the space dual to Y and $\Lambda^*\colon Y^* \to X^*$ is the operator conjugate with Λ.

THEOREM II. *If the elements $\widehat{x}$ and $\widehat{y}^*$ are admissible in problems (P_{II}) and (P_{II}^*) respectively and on these elements the equality between the values of these problems is attained,*

$$\langle x^*, \widehat{x}\rangle = \delta\|\widehat{y}^*\|_{Y^*}, \tag{II}$$

then $\widehat{x}$ is the solution to problem (P_{II}), and $\widehat{y}^*$ is the solution to problem (P_{II}^*). Moreover, the functional $\widehat{y}^*$ provides the optimal method in the corresponding recovery problem, i.e., $\langle x^*, x\rangle \approx \langle \widehat{y}^*, y\rangle$, $\|y - \Lambda x\|_Y \le \delta$, and $E(x^*, L, F) = \langle x^*, \widehat{x}\rangle$.

PROOF. Let x and y^* be arbitrary admissible elements in problems (P_{II}) and (P_{II}^*) respectively. Using elementary inequalities and (II) we get

$$\begin{aligned}\langle x^*, x\rangle &= \langle \Lambda^*\widehat{y}^*, x\rangle = \langle \widehat{y}^*, \Lambda x\rangle \le \|\widehat{y}^*\|_{Y^*}\|\Lambda x\|_Y \\ &\le \delta\|\widehat{y}^*\|_{Y^*} = \langle x^*, \widehat{x}\rangle = \langle \Lambda^* y^*, \widehat{x}\rangle \\ &= \langle y^*, \Lambda\widehat{x}\rangle \le \|y^*\|_{Y^*}\|\Lambda\widehat{x}\|_Y \le \delta\|y^*\|_{Y^*}.\end{aligned}$$

This implies that $\langle x^*, x\rangle \le \langle x^*, \widehat{x}\rangle$ for any element x admissible in (P_{II}) and $\delta\|\widehat{y}^*\|_{Y^*} \le \delta\|y^*\|_{Y^*}$ for any element y^* admissible in (P_{II}^*). Hence $\widehat{x}$ is a solution to problem (P_{II}) and $\widehat{y}^*$ is a solution to (P_{II}^*).

Next, using the fact that $\widehat{y}^*$ is an admissible element in (P_{II}^*), the equality between the values of the problems, and inequality (3), we obtain

$$
\begin{aligned}
E(x^*, L, F) &\le \sup_{x\in L,\, \|y-\Lambda x\|_Y\le\delta} |\langle x^*, x\rangle - \langle \widehat{y}^*, y\rangle| \\
&= \sup_{x\in L,\, \|z\|_Y\le\delta} |\langle x^*, x\rangle - \langle \widehat{y}^*, z+\Lambda x\rangle| \\
&= \sup_{x\in L,\, \|z\|_Y\le\delta} |\langle x^* - \Lambda^*\widehat{y}^*, x\rangle - \langle \widehat{y}^*, z\rangle| \\
&= \sup_{\|z\|_Y\le\delta} |\langle \widehat{y}^*, z\rangle| = \delta\|\widehat{y}^*\|_{Y^*} \\
&= \langle x^*, \widehat{x}\rangle \le E(x^*, L, F).
\end{aligned}
$$

Thus $\widehat{y}^*$ is an optimal method, which proves the last assertion of the theorem. □

Note that if $\dim L = n$, $x_1, \dots, x_n$ is a basis of L, and $\langle x^*, x_i\rangle = \alpha_i$, $1 \le i \le n$, then problem (P^*_{II}) is equivalent to

$$\delta\|y^*\|_{Y^*} \to \min, \quad \langle y^*, \Lambda x_i\rangle = \alpha_i, \quad i = 1, \dots, n. \qquad (P^*_{II_n})$$

Such a problem (minimization of the norm of a functional when its values on a fixed set of elements are known) is called the *problem of moments.*

Note also that the dual to this problem with respect to the perturbation

$$\delta\|y^*\|_{Y^*} \to \min, \qquad \langle y^*, \Lambda x_i\rangle = \alpha_i + \beta_i, \quad i = 1, \dots, n,$$

has the form

$$\sum_{i=1}^n \alpha_i\gamma_i \to \max, \qquad \left\|\sum_{i=1}^n \gamma_i\Lambda x_i\right\|_Y \le \delta, \quad \gamma = (\gamma_1, \dots, \gamma_n)^T \in \mathbb{R}^n.$$

This is problem (P_{II}) when $X = L = \mathbb{R}^n$, $x^* = (\alpha_1, \dots, \alpha_n) \in (\mathbb{R}^n)'$, and the role of Λ is played by the operator $\Lambda_1 \colon \mathbb{R}^n \to Y$, $\Lambda_1\gamma = \sum_{i=1}^n \gamma_i\Lambda x_i$ $(\gamma = (\gamma_1, \dots, \gamma_n))$.

Of course, results similar to the theorems just proved hold also for the problems stated above.

10.3. Specific problems.

1. *Polynomials least deviating from zero.* In problem (P_I), take $X = L_q(I)$ $(I = [-1, 1])$ for $1 \le q < \infty$ and $C(I)$ for $q = \infty$, let $L = \mathcal{P}_{n-1}$ be the n-dimensional space of algebraic polynomials $\xi(\cdot)$ of

degree at most $n-1$ ($\xi(t)=\sum_{k=0}^{n-1}\lambda_k t^k$), and let $x_0(\cdot)$ be the function $t\to t^n$. Problem (P_I) with these parameters has the form

$$\|x_0(\cdot)-\xi(\cdot)\|_{X_q}\to\min,\quad \xi(\cdot)\in\mathcal{P}_{n-1},\tag{P_{nq}}$$

where $X_q=L_q(I)$ for $1\le q<\infty$ and $X_q=C(I)$ for $q=\infty$.

Let $\widehat{\xi}_{nq}(\cdot)$ be a solution of this problem (which exists since the space $\mathcal{P}_{n-1}$ is finite-dimensional). Set $T_{nq}(t)=t^n-\widehat{\xi}_{nq}(t)$. This polynomial (of degree n, with leading coefficient 1, and the minimal norm in the metric of $L_q(I)$) is called the *polynomial least deviating from zero in the metric of* $L_q(I)$.

The dual problem to (P_{nq}) for $1\le q<\infty$ has the form

$$\begin{gathered}\int_I t^n x(t)\,dt\to\max,\\ \int_I t^k x(t)\,dt=0,\quad k=0,1,\dots,n-1,\quad x(\cdot)\in BL_{q'}(I),\end{gathered}\tag{P^*_{nq}}$$

and for $q=\infty$ the form

$$\begin{gathered}\int_I t^n\,d\mu(t)\to\max,\\ \int_I t^k\,d\mu(t)=0,\quad k=0,1,\dots,n-1,\quad \mu(\cdot)\in B\operatorname{Var}(I),\end{gathered}\tag{$P^*_{n\infty}$}$$

where $\operatorname{Var}(I)$ is the space of functions $\mu(\cdot)$ of bounded variation on I (equal to zero at zero and right continuous on $(0,1]$) with norm the total variation $\operatorname{var}(\mu(\cdot),I)$ of $\mu(\cdot)$ on I.

For $1\le q<\infty$ the optimal recovery problem corresponding to (P^*_{nq}) consists in optimal recovery of the functional $m_{nq'}\colon x(\cdot)\to\int_I t^n x(t)\,dt$ (the nth moment of $x(\cdot)$) on the class $BL_{q'}(I)$ from the lower order moments, i.e., based on the information $F_{n\,\mathrm{mom}\,q'}\colon L_{q'}(I)\to\mathbb{R}^n$, $F_{n\,\mathrm{mom}\,q'}x(\cdot)=(m_{0q'},\dots,m_{n-1,q'})$, where $m_{kq'}=\int_I t^k x(t)\,dt$, $k=0,1,\dots,n-1$.

A similar problem can be stated for $q=\infty$.

The complete solution of problem (P_{nq}) has been found in the cases where $q=2,\infty,1$.

THEOREM 1. *Let $n\in\mathbb{N}$. Then for $1\le q<\infty$ the polynomial $T_{nq}(\cdot)$ is completely characterized by the relation*

$$\int_I |T_{nq}(t)|^{q-1}\operatorname{sign}T_{nq}(t)\xi(t)\,dt=0,\quad \forall\xi(\cdot)\in\mathcal{P}_{n-1}.\tag{O_q}$$

If $1 \le q \le \infty$ *and* $X_q^* = L_{q'}(I)$ *for* $1 \le q < \infty$ *and* $X_q^* = \mathrm{Var}(I)$ *for* $q = \infty$, *then* $E(m_{nq'}, BX_q^*, F_{n\,\mathrm{mom}\,q'}) = d(x_0(\cdot), \mathcal{P}_{n-1}, X_q) = \|T_{nq}(\cdot)\|_{X_q}$ *and* $m_{nq'} \approx \sum_{k=0}^{n-1} \lambda_{knq'} m_{kq'}$, *where the* $\lambda_{knq'}$, $0 \le k \le n-1$, *are the coefficients of the best approximation polynomial* $\widehat{\xi}_{nq'}(\cdot)$.

Furthermore,

$$T_{n2}(t) = \frac{n!}{(2n)!}\frac{d^n}{dt^n}(t^2-1)^n,$$

$$E(m_{n2}, BL_2(I), F_{n\,\mathrm{mom}\,2}) = \frac{2^n(n!)^2}{(2n)!}\sqrt{\frac{2}{2n+1}},$$

$$T_{n\infty}(t) = 2^{-(n-1)}\cos n \arccos t,$$

$$E(m_{n\infty}, B\,\mathrm{Var}(I), F_{n\,\mathrm{mom}\,\infty}) = 2^{-(n-1)},$$

$$T_{n1}(t) = \frac{1}{n+1}\dot{T}_{n+1,\infty}(t) = \frac{1}{2^n(n+1)}\frac{\sin n \arccos t}{\sqrt{1-t^2}},$$

$$E(m_{n1}, BL_\infty(I), F_{n\,\mathrm{mom}\,1}) = 2^{-(n-1)}.$$

REMARK 1. The polynomials $P_n(t) = \frac{1}{2^n n!}\frac{d^n}{dt^n}(t^2-1)^n$ proportional to $T_{n2}(\cdot)$ are called the Legendre polynomials, and the formulas defining them are called the Rodrigues formulas. The polynomials $\sqrt{(2n+1)/2}P_n(\cdot)$ form an orthonormal system in $L_2(I)$.

The polynomials $T_n(t) = \cos n \arccos t$ are called the Chebyshev polynomials of the first kind, and $T_{n1}(\cdot)$ the Chebyshev polynomials of the second kind. Their explicit expression was obtained by A. N. Korkin and E. I. Zolotarev in the 1880s.

PROOF. Problem (P_{nq}) for $1 \le q < \infty$ is the minimization problem of a smooth convex function of n variables (the coefficients of the polynomial $\widehat{\xi}(\cdot)$) on the entire space; hence for a point $(\lambda_{0nq'}, \dots, \lambda_{n-1,nq'})$ (the set of coefficients of $\widehat{\xi}_{nq'}(\cdot)$) to be its solution it is necessary and sufficient that

$$\int_I |T_{nq}(t)|^{q-1}\operatorname{sign} T_{nq}(t) t^k\, dt = 0, \quad k = 0, 1, \dots, n-1, \qquad (i)$$

which is, obviously, equivalent to (O_q).

Set $\widehat{x}_{nq}(\cdot) = \|T_{nq}(\cdot)\|_{L_q(I)}^{1-q} |T_{nq}(\cdot)|^{q-1} \operatorname{sign} T_{nq}(\cdot)$. We see from this expression and (i) that $\widehat{x}(\cdot)$ is an admissible function in (P_{nq}^*) and

$$\begin{aligned}\int_I t^n \widehat{x}_{nq}(t)\,dt &= \int_I (t^n - \widehat{\xi}_{nq}(t))\widehat{x}_{nq}(t)\,dt \\ &= \int_I T_{nq}(t)\widehat{x}_{nq}(t)\,dt = \|T_{nq}(\cdot)\|_{L_q(I)} \\ &= \|x_0(\cdot) - \widehat{\xi}_{nq}(\cdot)\|_{L_q(I)} = d(x_0(\cdot), \mathcal{P}_{n-1}, L_q(I)).\end{aligned} \tag{ii}$$

Thus the values of problems (P_{nq}) and (P_{nq}^*) are equal; hence $\widehat{x}_{nq}(\cdot)$ is a solution to problem (P_{nq}^*). Then Theorem I implies the expression for the optimal recovery method of the nth moment from the lower order moments.

Now we proceed to the determination of explicit expressions for the polynomials least deviating from zero in the cases of $q = 2, \infty$, and 1.

Case $q = 2$. Set $p(t) = \frac{1}{(n-1)!}\int_{-1}^t (t-\tau)^{n-1} T_{n2}(\tau)d\tau$. It is not hard to see that $p(\cdot) \in \mathcal{P}_{2n}$, $p^{(n)}(\cdot) = T_{n2}(\cdot)$, and $p^{(k)}(-1) = 0$, $0 \le k \le n-1$. Relation (i) for $q = 2$ implies that $p^{(k)}(1) = 0$, $0 \le k \le n-1$. Thus $p(t) = c(t^2-1)^n$; hence $T_{n2}(t) = c((t^2-1)^n)^{(n)}$, where the coefficient c is chosen so that the leading coefficient is equal to one. Therefore $T_{n2}(t) = (n!/(2n)!)((t^2-1)^n)^{(n)}$.

Case $q = \infty$. Let $\widehat{d\mu}(\cdot)$ be an admissible measure in problem $(P_{n\infty}^*)$. Assume that the values of problems $(P_{n\infty})$ and $(P_{n\infty}^*)$ are equal, which, due to $d\widehat{\mu}(\cdot) \in \mathcal{P}_{n-1}^{\perp}$, may be written as

$$\int_I T_{n\infty}(t)d\widehat{\mu}(t) = \|T_{n\infty}(\cdot)\|_{C(I)}. \tag{iii}$$

Consider the set $M = \{t \in I \mid |T_{n\infty}(t)| = \|T_{n\infty}(\cdot)\|_{C(I)}\}$. It is clear that it consists of at most $n+1$ points. We will show that the measure $d\widehat{\mu}(\cdot)$ is concentrated on M; i.e., the function $\widehat{\mu}(\cdot)$ is a step function on I with discontinuities at the points of M. To this end it suffices to check that the total variation of $\widehat{\mu}(\cdot)$ on any interval I_0 in I equals zero. Assume the contrary, i.e., that the total variation of $\widehat{\mu}(\cdot)$ on some interval I_0 is different from zero. Then we arrive at a contradiction:

$$\begin{aligned}
&\|T_{n\infty}(\cdot)\|_{C(I)} \\
&= \int_I T_{n\infty}(t)d\widehat{\mu}(t) = \int_{I_0} T_{n\infty}(t)d\widehat{\mu}(t) + \int_{I\setminus I_0} T_{n\infty}(t)d\widehat{\mu}(t) \\
&\le \max_{t\in I_0} |T_{n\infty}(t)| \operatorname{var}(\widehat{\mu}(\cdot), I_0) + \|T_{n\infty}(\cdot)\|_{C(I)} \operatorname{var}(\widehat{\mu}(\cdot), I\setminus I_0) \\
&< \|T_{n\infty}(\cdot)\|_{C(I)}(\operatorname{var}(\widehat{\mu}(\cdot), I_0) + \operatorname{var}(\widehat{\mu}(\cdot), I\setminus I_0)) \\
&= \|T_{n\infty}(\cdot)\|_{C(I)} \operatorname{var}(\widehat{\mu}(\cdot), I) \le \|T_{n\infty}(\cdot)\|_{C(I)}.
\end{aligned}$$

Let $-1 \le \tau_1 < \tau_2 < \cdots < \tau_r \le 1$, $r \le n+1$, be the points of M. Now the condition $d\widehat{\mu}(\cdot) \in \mathcal{P}_{n-1}^{\perp}$ may be rewritten as

$$\sum_{k=1}^{r} \widehat{\mu}_k \xi(\tau_k) = 0, \quad \forall \xi(\cdot) \in \mathcal{P}_{n-1}, \tag{iv}$$

where $\sum_{k=1}^{r} |\widehat{\mu}_k| = 1$.

We will show that $r = n+1$. Indeed, assume that $r \le n$. Observe first of all that $r \ge 2$, since if $r = 1$, then putting a constant different from zero into (iv) we get $\widehat{\mu}_1 = 0$ in contradiction with $|\widehat{\mu}_1| = 1$. The numbers $\widehat{\mu}_i$, $1 \le i \le r$, cannot all be equal to zero. Let, for example, $\widehat{\mu}_r \ne 0$. Consider the polynomial $p(t) = \Pi_{j=1}^{r-1}(t-\tau_j)$. Putting it into (iv) we obtain that $\widehat{\mu}_r = 0$, which contradicts our assumption. Thus $r = n+1$.

Now we note that necessarily $\tau_1 = -1$ and $\tau_{n+1} = 1$, because if at least one of these equalities failed, the polynomial $\dot{T}_{n\infty}(\cdot)$ of degree $n-1$ would have n roots (obviously, the derivative of $T_{n\infty}(\cdot)$ vanishes at the points of M lying inside I) and hence it would be identically equal to zero, which is impossible. Therefore the polynomials $(1-t^2)\dot{T}^2_{n\infty}(t)$ and $a^2 - T^2_{n\infty}(t)$ (where $a = \|T_{n\infty}(\cdot)\|_{C(I)}$) having the same degree $2n$ and the same roots (of multiplicity 1 at ± 1 and 2 at τ_i, $2 \le i \le n$) differ only by a constant factor. Taking this factor to be n^2 to make equal the leading coefficients, we see that these polynomials are equal; hence the desired polynomial must satisfy the equation $(1-t^2)\dot{x}^2 = n^2(a^2 - x^2)$. Integrating this equation and choosing the constant so that the leading coefficient is equal to one, we obtain that $T_{n\infty}(t) = 2^{-(n-1)} \cos n \arccos t$. The points τ_k at which the polynomial alternately takes its maximal and minimal values are $\tau_k = \cos \frac{(n+1-k)\pi}{n}$, $k = 1, \dots, n+1$.

Now we present the $\widehat{\mu}_k$ satisfying (iv). Set

$$\widehat{\mu}_{n+1} = \left(1 + \sum_{j=1}^{n} |\zeta_j(1)|\right)^{-1}, \qquad \widehat{\mu}_k = -\widehat{\mu}_{n+1}\zeta_k(1),\ k = 1, \dots, n,$$

where

$$\zeta_j(\cdot) \in \mathcal{P}_{n-1} \quad \text{and} \quad \zeta_j(\tau_k) = \delta_{jk},\ j, k = 1, \dots, n.$$

Relation (iv) with these coefficients $\widehat{\mu}_k$ is obviously fulfilled for the polynomials $\zeta_j(\cdot)$, $1 \le j \le n$. But these polynomials form a basis in $\mathcal{P}_{n-1}$; hence (iv) with such constructed coefficients holds true. Next, clearly $\sum_{k=1}^{n+1} |\widehat{\mu}_k| = 1$; therefore the function $\widehat{\mu}(\cdot)$ constant on the intervals (τ_k, τ_{k+1}), $k = 1, \dots, n$, and such that $\widehat{\mu}(-1) = 0$ and $\widehat{\mu}(\tau_k) = \sum_{i=1}^{k} \widehat{\mu}_i$, $k = 2, \dots, n+1$, induces the measure $d\widehat{\mu}(\cdot)$ which is admissible in problem $(P^*_{n\infty})$. It is not hard to see that $\operatorname{sign} \zeta_k(\tau_1) = (-1)^{n+k}$, $k = 1, \dots, n$; i.e., $\operatorname{sign} \widehat{\mu}_k = \operatorname{sign} T_{n\infty}(\tau_k)$, $k = 1, \dots, n$. Then $\sum_{k=1}^{n+1} \widehat{\mu}_k T_{n\infty}(\tau_k) = \sum_{k=1}^{n+1} |\widehat{\mu}_k| \|T_{n\infty}(\cdot)\|_{C(I)} = \|T_{n\infty}(\cdot)\|_{C(I)}$; i.e., the equality (iii) between the values of the two problems holds. The rest follows from Theorem I.

Case $q = 1$. Let $\widehat{x}_{n1}(\cdot)$ be an admissible function in (P^*_{n1}) and let the values of the problems (P_{n1}) and (P^*_{n1}) be equal:

$$\int_I T_{n1}(t)\widehat{x}_{n1}(t)\, dt = \|T_{n1}(\cdot)\|_{L_1(I)}. \tag{v}$$

This implies that the function $\widehat{x}_{n1}(\cdot)$ is proportional to $\operatorname{sign} T_{n1}(\cdot)$; hence $\widehat{x}_{n1}(\cdot) \in \mathcal{P}^{\perp}_{n-1}$ is equivalent to

$$\int_I t^k \operatorname{sign} T_{n1}(t)\, dt = 0, \quad k = 0, 1, \dots, n-1.$$

In the previous part of the proof we have shown that the measure $d\widehat{\mu}(\cdot)$ induced by the function $\widehat{\mu}(\cdot) = (2n)^{-1}\dot{T}_{n\infty}(\cdot)$ is orthogonal to polynomials of degree not exceeding $n-1$. Hence replacing n by $n+1$ we have

$$\int_I t^k d \operatorname{sign} \dot{T}_{n+1,\infty}(t) = 0, \quad k = 0, 1, \dots, n.$$

Integrating by parts we obtain

$$\int_I t^k \operatorname{sign} \dot{T}_{n+1,\infty}(t)\, dt = 0, \quad k = 0, 1, \dots, n-1.$$

Set $T_{n1}(\cdot) = (n+1)^{-1}\dot{T}_{n+1,\infty}(\cdot)$ (the factor $(n+1)^{-1}$ makes the leading coefficient equal to one) and $\widehat{x}_{n1}(\cdot) = \|T_{n1}(\cdot)\|_{L_1}^{-1} T_{n1}(\cdot)$. Then obviously (v) holds, and the desired assertion follows from Theorem I.$\square$

2. *Chebyshev's problem on extrapolation of polynomials.* This problem (already mentioned above) is formally stated as follows:

$$x(\tau) \to \max, \quad \|x(\cdot)\|_{C(I)} \le \delta, \quad x(\cdot) \in \mathcal{P}_n, \tag{i}$$

where $|\tau| > 1$, $\delta > 0$, $I = [-1, 1]$, and $\mathcal{P}_n$ is the space of algebraic polynomials of degree not exceeding n.

This problem may be considered as the one associated with the optimal recovery problem of the linear functional $x(\cdot) \to x(\tau)$ (defined on the vector space of all continuous functions on $\mathbb{R}$) on the class $\mathcal{P}_n$ based on the following information: for each polynomial $x(\cdot) \in \mathcal{P}_n$ we know $y(\cdot) \in C(I)$ which differs from $x(\cdot)$ by at most δ in the metric of $C(I)$; i.e., the information operator is the multi-valued mapping $F_{\delta C(I)}: \mathcal{P}_n \to C(I)$, $F_{\delta C(I)} x(\cdot) = x(\cdot)|_I + \delta BC(I)$.

THEOREM 2. *The function $\widehat{x}(\cdot) = (\operatorname{sign}\tau)^n \delta T_n(\cdot)$, where $T_n(t) =$ $\cos n \arccos t$ is the Chebyshev polynomial, is a solution to the problem (i). The functional $y(\cdot) \to \sum_{k=1}^{n+1} \zeta_k(\tau) y(\tau_k)$ on $C(I)$, where $\zeta_j(\cdot) \in \mathcal{P}_n$, $\zeta_j(\tau_k) = \delta_{jk}$, $j, k = 1, \dots, n+1$, and $\tau_k = \cos((n-k+1)\pi/n)$, $k = 1, \dots, n+1$, is the optimal method in the corresponding recovery problem (i.e., the method associates with $y(\cdot)$ the value at the point τ of the Lagrange polynomial interpolating $y(\cdot)$ at the alternation points of the Chebyshev polynomial $T_n(\cdot)$).*

PROOF. According to subsection *II* the dual problem to (i) has the form

$$\delta \operatorname{var}(\mu(\cdot), I) \to \max, \qquad \int_I x(t)\, d\mu(t) = x(\tau), \quad x(\cdot) \in \mathcal{P}_n. \tag{ii}$$

Let $\widehat{x}(\cdot)$ and $\widehat{\mu}(\cdot)$ be admissible elements in the problems (i) and (ii) respectively, and let the equality between the values of these problems be attained on them:

$$\delta \operatorname{var}(\widehat{\mu}(\cdot), I) = \widehat{x}(\tau). \tag{iii}$$

Consider the set $M = \{t \in I \mid |\widehat{x}(t)| = \delta\}$. It is not empty, since (ii) and (i) imply that $\delta \operatorname{var}(\widehat{\mu}(\cdot), I) = \widehat{x}(\tau) = \int_I \widehat{x}(t)\, d\widehat{\mu}(t) \le$ $\|\widehat{x}(\cdot)\|_{C(I)} \operatorname{var}(\mu(\cdot), I)$, i.e., $\|\widehat{x}(\cdot)\|_{C(I)} \ge \delta$, so that $\|\widehat{x}(\cdot)\|_{C(I)} = \delta$. It is clear that M consists of at most $n+1$ points $-1 \le \tau_1 < \cdots < \tau_r \le 1$, $r \le n+1$.

Similarly to the previous example one can check that the measure $d\widehat{\mu}(\cdot)$ is supported on the set M, and it is positive (negative) at the points where $\widehat{x}(\cdot)$ is positive (negative). Then its admissibility in the problem (ii) implies that

$$\sum_{k=1}^{r} \widehat{\mu}_k x(\tau_k) = x(\tau), \quad \forall x(\cdot) \in \mathcal{P}_n. \tag{iv}$$

Let us show that $r = n + 1$. Indeed, if $r \leq n$, then substituting the polynomial $p(t) = \prod_{k=1}^{r}(t - \tau_k)$ into (iv) we arrive at a contradiction: $0 = p(\tau) \neq 0$.

Since $r = n + 1$, the function $\widehat{x}(\cdot)$ equals, up to a constant factor, the Chebyshev polynomial $T_n(\cdot)$ (as was shown in the previous problem). We choose this factor so that, on the one hand, $\|\widehat{x}(\cdot)\|_{C(I)} = \delta$ and, on the other hand, $\widehat{x}(\tau) > 0$ (the value of problem (i) is obviously positive). Then using the fact that $T_n(t) > 0$ for $t \geq 1$ and $(-1)^n T_n(t) > 0$ for $t \leq 1$, we set $\widehat{x}(\cdot) = (\operatorname{sign} \tau)^n \delta T_n(\cdot)$. It is clear that $\widehat{x}(\cdot)$ is an admissible function in problem (i).

Set $\widehat{\mu}_k = \zeta_k(\tau)$, where $\zeta_k(\cdot)$, $k = 1, \ldots, n+1$, are the functions defined in the statement of the theorem. These functions obviously fulfill relation (iv), and since they form a basis in $\mathcal{P}_n$, this implies (iv) with the $\widehat{\mu}_k$, $1 \leq k \leq n+1$, so defined. Therefore the function $\widehat{\mu}(\cdot)$ which is constant on the intervals (τ_k, τ_{k+1}), $k = 1, \ldots, n$, and is such that $\widehat{\mu}(-1) = 0$ and $\widehat{\mu}(\tau_k) = \sum_{i=1}^{k} \widehat{\mu}_i$, $k = 2, \ldots, n+1$, induces the measure $d\mu(\cdot)$ admissible in problem (ii). Next, it is easy to see that $\operatorname{sign} \widehat{\mu}_k = (\operatorname{sign} \tau)^n (-1)^{n+k+1}$, $k = 1, \ldots, n+1$. Recall that $T_n(\tau_k) = (-1)^{n+k+1}$. Then we get

$$\begin{aligned}
\delta \operatorname{var}(\widehat{\mu}(\cdot), I) &= \delta \sum_{k=1}^{n+1} |\widehat{\mu}_k| = \delta (\operatorname{sign} \tau)^n \sum_{k=1}^{n+1} \widehat{\mu}_k (-1)^{n+k+1} \\
&= \delta (\operatorname{sign} \tau)^n \sum_{k=1}^{n+1} \widehat{\mu}_k T_n(\tau_k) \\
&= \delta (\operatorname{sign} \tau) T_n(\tau) = \widehat{x}(\tau);
\end{aligned}$$

i.e., the function $\widehat{x}(\cdot)$ and the measure $d\widehat{\mu}(\cdot)$ constructed above fulfill equality (iii), hence they provide the solutions of the corresponding problems. The form of the optimal recovery method follows immediately from the general result. $\square$

3. *The problem by A. Markov and B. Markov on inequalities for derivatives of algebraic polynomials.* Let $n \in \mathbb{N}$, $1 \le k < n$, and $\tau \in \mathbb{R}$. The problem by A. Markov and B. Markov is as follows: among algebraic polynomials of degree at most n and not exceeding in absolute value a fixed $\delta > 0$ on the interval $[-1,1]$, find the one attaining the maximal value of the kth derivative at a fixed point τ. In other words, we are interested in the extremal problem

$$x^{(k)}(\tau) \to \max, \qquad \|x(\cdot)\|_{C([-1,1])} \le \delta, \quad x(\cdot) \in \mathcal{P}_n. \tag{i}$$

This problem may be viewed as the one associated with the optimal recovery problem of the linear functional $x(\cdot) \to x^{(k)}(\tau)$ on the same class and based on the same information as in the previous example.

Before stating the next theorem we have to make some explanations. A continuous function $x(\cdot)$ on an interval $[a,b]$ is said to have the k-alternance if there exist points $a \le \tau_1 < \cdots < \tau_k \le b$ such that $|x(\tau_i)| = \|x(\cdot)\|_{C([-1,1])}$, $i = 1, \dots, k$, and $x(\tau_i)x(\tau_{i+1}) < 0$, $i = 1, \dots, k-1$.

A polynomial of degree n having the $(n+1)$-alternance on the interval $[-1,1]$ is proportional to Chebyshev's polynomial $T_n(\cdot)$ (this has actually been proved in Theorem 1). Besides that, there is a one-parameter family of polynomials of degree n having the n-alternance. These are called Zolotarev's polynomials (see Akhiezer (1965)).

THEOREM 3. *Let $n \in \mathbb{N}$, $1 \le k < n$, and $\tau \in \mathbb{R}$. Then:*

(a) *If $|\tau| > 1$ then the solution to the problem (i) is either $\delta T_n(\cdot)$ or $-\delta T_n(\cdot)$. In this case $E(x^{(k)}(\tau), \mathcal{P}_n, F_{\delta C([-1,1])}) = \delta |T_n^{(k)}(\tau)|$, and the functional $y(\cdot) \to \sum_{j=1}^{n+1} \zeta_j^k(\tau) y(\cos((n-j+1)\pi/n))$ on $C([-1,1])$, where the $\zeta_j(\cdot)$ are the same as in Theorem 2, is the optimal recovery method.*

(b) *If $|\tau| \le 1$ then the solution to the problem (i) is either $\pm\delta T_n(\cdot)$ or one of Zolotarev's polynomials. Here $E(x^{(k)}(\tau), \mathcal{P}_n, F_{\delta C([-1,1])}) = \delta |P_{nk\tau}^{(k)}(\tau)|$, where $P_{nk\tau}(\cdot)$ is either Chebyshev's or Zolotarev's polynomial, and the functional $y(\cdot) \to \sum_{j=1}^{r} \zeta_j^k(\tau) y(\tau_{nkj\tau})$ on $C([-1,1])$, where $r = n+1$ or n respectively and the $\tau_{nkj\tau}$ are the alternance points of $P_{nk\tau}(\cdot)$, is the optimal recovery method.*

PROOF. The problem dual to (i) is

$$\delta \operatorname{var}(\mu(\cdot), I) \to \max, \qquad \int_I x(t)d\mu(t) = x^{(k)}(\tau), \quad x(\cdot) \in \mathcal{P}_n. \quad (ii)$$

Let $\widehat{x}(\cdot)$ and $\widehat{\mu}(\cdot)$ be admissible elements in the problems (i) and (ii) respectively, and let the equality between the values of these problems be attained on them:

$$\delta \operatorname{var}(\widehat{\mu}(\cdot), I) = \widehat{x}^{(k)}(\tau). \qquad (iii)$$

As before, we can check that the measure $d\widehat{\mu}(\cdot)$ is supported on the set of $t \in I$ such that $|\widehat{x}(t)| = \delta$, and it is positive (negative) where $\widehat{x}(\cdot)$ is positive (negative). Therefore its admissibility in the problem (ii) means that

$$\sum_{k=1}^{r} \widehat{\mu}_k x(\tau_k) = x^{(k)}(\tau), \quad \forall x(\cdot) \in \mathcal{P}_n. \qquad (iv)$$

(a) Let us show that in this case $r = n+1$. Indeed, assume that $r \le n$. Consider the polynomial $p_1(t) = \prod_{i=1}^{r}(t-\tau_i)$. By the Rolle theorem $p_1^{(k)}(\cdot)$ may have roots only in $[-1,1]$; hence $p_1^{(k)}(\tau) \neq 0$. But putting $p_1(\cdot)$ into (iv), we obtain that $p_1^{(k)}(\tau) = 0$. This contradiction shows that $r = n+1$. As in the proof of Theorem 1, this implies that $\widehat{x}(\cdot)$ is either $\delta T_n(\cdot)$ or $-\delta T_n(\cdot)$.

(b) Let us show that $r \ge n$. If $n = 1$, this is obvious. Let $n \ge 2$ and assume that $r \le n-1$. Consider the polynomial $p_2(t) = \prod_{i=1}^{n-1}(t-\tau_i)$ (adding the points $\tau_{r+1}, \dots, \tau_{n-1}$ if necessary). Putting it into (iv) we obtain that $p_2^{(k)}(\tau) = 0$. For $k = n-1$ this is impossible. Let $k < n-1$. Set $p_3(t) = (t-\tau)p_2(t)$. Now putting this polynomial into (iv) we obtain that $p_3^{(k)}(\tau) = 0$. But $p_3^{(k)}(\tau) = kp_2^{(k-1)}(\tau)$; hence $p_2^{(k-1)}(\tau) = p_2^{(k)}(\tau) = 0$. Thus the polynomial $p_2^{(k-1)}(\cdot)$ of degree $n-k$ has at least $n-k+1$ roots ($n-k$ roots by the Rolle theorem and a multiple root at the point τ), which is impossible. Hence $r = n$ or $n+1$. If $r = n+1$, then, as we proved, $\widehat{x}(\cdot)$ equals either $\delta T_n(\cdot)$ or $-\delta T_n(\cdot)$. Let $r = n$. Let us check that the points τ_i, $i = 1, \dots, n$ (one may assume that $\tau_1 < \dots < \tau_n$) form the n-alternance for $\widehat{x}(\cdot)$. Assume that $\widehat{x}(\tau_j) = \widehat{x}(\tau_{j+1})$ for some $1 \le j \le n-1$. It is clear that there exists a nontrivial polynomial $p_4(\cdot) \in \mathcal{P}_n$ such that $p_4(\tau_i) = 0$, $i = 1, \dots, j-1, j+2, \dots, n$, and $p_4^{(k-1)}(\tau) = p_4^{(k)}(\tau) = 0$. Putting $p_4(\cdot)$ into (iv) we arrive at a contradiction: $\widehat{\mu}_k + \widehat{\mu}_{k+1} = 0$. Thus the points τ_i, $i = 1, \dots, n$, form the n-alternance, which means that $\widehat{x}(\cdot)$ is one of Zolotarev's polynomials.

The errors of optimal recovery and the form of the optimal methods follow from the general results. □

4. *Bernstein's inequality for trigonometric polynomials.* The problem of finding the best constant in Bernstein's inequality

$$\|\dot{x}(\cdot)\|_{C(\mathbb{T})} \leq K\|x(\cdot)\|_{C(\mathbb{T})}, \quad x(\cdot) \in \mathcal{T}_n, \tag{i}$$

where $\mathcal{T}_n$ is the space of trigonometric polynomials of degree not exceeding n, is equivalent (due to translation invariance of the norm) to the following extremal problem:

$$\dot{x}(0) \to \max, \qquad \|x(\cdot)\|_{C(\mathbb{T})} \leq \delta, \quad x(\cdot) \in \mathcal{T}_n. \tag{ii}$$

This problem is associated with the optimal recovery problem of the functional $x(\cdot) \to \dot{x}(0)$ on the class $\mathcal{T}_n$ based on the same information as in the previous examples.

Arguing as in the previous example, we get, in particular, an analog of relation (iv):

$$\sum_{k=1}^{r} \widehat{\mu}_k x(\tau_k) = \dot{x}(0), \quad \forall x(\cdot) \in \mathcal{T}_n, \tag{iii}$$

with $r \leq 2n$ (since a trigonometric polynomial cannot have more than $2n$ extrema). We will show that $r = 2n$. Indeed, suppose $r \leq 2n - 1$. Let us add to the points $\{\tau_i\}_{i=1}^r$ the point 0 and perhaps some more points to make their total number $2n$. There exists a nonzero polynomial with roots at these points. Then its derivative necessary differs from zero at the point 0, which contradicts (iii).

If $\widehat{x}(\cdot)$ is an extremal function in (ii), then the polynomials $\dot{\widehat{x}}^2(\cdot)$ and $\delta^2 - \widehat{x}^2(\cdot)$, which belong to $\mathcal{T}_n$, have roots of multiplicity two at the points $\tau_1, \dots, \tau_{2n}$; hence they are proportional. Comparing the coefficients of the leading terms, we obtain that the solution must satisfy the equation $\dot{x}^2(\cdot) = n^2(\delta^2 - x^2(\cdot))$. Integrating this equation we find that $\widehat{x}(t) = \delta \sin nt$.

This immediately implies that the best constant in Bernstein's inequality (i) is equal to n.

In the present periodic case the constants in the relation similar to (iv) of the previous example can be easily evaluated, and we obtain the identity

$$\dot{x}(0) = \frac{1}{2n}\sum_{k=1}^{2n} x(\tau_k)(-1)^{k+1}\frac{1}{2\sin^2\tau_k/2},$$
$$\tau_k = \frac{(2k-1)\pi}{2n}, \quad k = 1, \dots, 2n,$$

referred to as the *M. Riesz identity.*

The assertions just obtained imply the following:

THEOREM 4. *Let $n \in \mathbb{N}$ and $\delta > 0$. Then:*

(a) $E(\dot{x}(0), \mathcal{T}_n, F_{\delta C(\mathbb{T})}) = \delta n$;

(b) *the linear functional*

$$y(\cdot) \to \frac{1}{2n}\sum_{k=1}^{2n}(-1)^{k+1}\frac{1}{2\sin^2\tau_k/2}y(\tau_k),$$
$$\tau_k = \frac{(2k-1)\pi}{2n}, \quad k = 1, \dots, 2n,$$

is the optimal recovery method;

(c) *the following exact inequality (Bernstein's inequality) holds:*

$$\|\dot{x}(\cdot)\|_{C(\mathbb{T})} \le n\|x(\cdot)\|_{C(\mathbb{T})}, \quad x(\cdot) \in \mathcal{T}_n.$$

5. *Recovery of an integral of a Sobolev class using boundary data.* Let, as before, $I = [-1, 1]$, $1 \le p \le \infty$, $n \in \mathbb{N}$, and let $W_p^n(I)$ be a Sobolev class of functions $x(\cdot)$ on I with absolute continuous $(n-1)$th derivative and $\|x^{(n)}(\cdot)\|_{L_p(I)} \le 1$. We wish to recover the integral $J(x(\cdot)) = \int_I x(t)dt$ on the class $W_p^n(I)$ based on the information: for each $x(\cdot) \in W_p^n(I)$ the values of $x^{(k)}(\pm 1)$, $0 \le k \le n-1$, are known precisely. Thus $X = L_1(I)$, $C = W_p^n(I)$, $Y = \mathbb{R}^{2n}$, $Fx(\cdot) = F_\partial x(\cdot) = \{x^{(k)}(\pm 1)\}_{k=0}^{n-1}$, and $\langle x', x(\cdot)\rangle = J(x(\cdot))$.

THEOREM 5. *Let $1 \le p < \infty$, $n \in \mathbb{N}$, and let $T_{np'}(\cdot)$ be the polynomial of degree n least deviating from zero in the norm of $L_{p'}$. Then*

(a) $E(J, W_p^n(I), F_\partial) = \frac{1}{n!}\|T_{np'}(\cdot)\|_{L_{p'}(I)}$;

(*b*) *the functional*

$$x(\cdot) \to R(x(\cdot))$$
$$= \frac{(-1)^{n-1}}{n!} \sum_{j=0}^{n-1} (-1)^j \left(T_{np'}^{(j)}(1) x^{(n-j-1)}(1) - T_{np'}^{(j)}(-1) x^{(n-j-1)}(-1) \right)$$

is the optimal recovery method.

PROOF. The associated problem has the form

$$\int_I x(t)dt \to \max, \qquad x^{(k)}(\pm 1) = 0, \quad 0 \le k \le n-1, \quad x(\cdot) \in W_p^n(I). \tag{i}$$

Its value is identical to the value of the following problem:

$$\frac{1}{n!} \int_I t^n u(t)dt \to \max,$$
$$\int_I t^k u(t)dt = 0, \quad 0 \le k \le n-1, \ u(\cdot) \in BL_p(I). \tag{ii}$$

Indeed, the formula

$$x(t) = \frac{1}{(n-1)!} \int_{-1}^{t} (t-\tau)^{n-1} u(\tau) d\tau$$

establishes a one-to-one correspondence between admissible elements in (i) and (ii). Integrating this formula we obtain the equality between the functionals maximized in (i) and (ii). But by Theorem 1 the value of the problem (ii) is equal to $(n!)^{-1} \|T_{np'}(\cdot)\|_{L_{p'}(I)}$. Next, integration by parts yields the relation

$$\frac{1}{n!} \int_I T_{np'}(t) x^{(n)}(t)dt = (-1)^n n! (J(x(\cdot)) - R(x(\cdot)),$$

whence, using Hölder's inequality and the fact that the value of the associated problem does not exceed the optimal recovery error, we obtain for all $x(\cdot) \in W_p^n(I)$

$$E(J, W_p^n(I), F_\partial) \le |J(x(\cdot)) - R(x(\cdot))| = \frac{1}{n!} \left| \int_I T_{np'}(t) x^{(n)}(t)dt \right|$$
$$\le \frac{1}{n!} \|T_{np'}(\cdot)\|_{L_{p'}(I)} \le E(J, W_p^n(I), F_\partial).$$

This implies both statements of the theorem. □

In particular, if $n = 1$, then $T_{1p'}(t) = t$, $E(J, W_p^1(I), F_\partial) = \|T_{1p'}(\cdot)\|_{L_{p'}(I)} = (\frac{2}{p'+1})^{1/p'}$ and $R(x(\cdot)) = x(1) + x(-1)$.

If $n = 2$ and $p = 1$, then $T_{2\infty}(t) = t^2 - \frac{1}{2}$, $E(J, W_1^2(I), F_\partial) = \|T_{2\infty}(\cdot)\|_{C(I)} = \frac{1}{2}$, and $R(x(\cdot)) = x(1) + x(-1) - \frac{1}{4}(\dot{x}(1) - \dot{x}(-1))$.

6. *Recovery of the integral using boundary data on the plane.* We discuss here a two-dimensional analog of the previous problems (the case of larger dimension is treated similarly).

Let D be a bounded domain on the plane with piece-wise smooth boundary ∂D. Denote by $W_2^1(D)$ the class of functions $x(\cdot)$ such that $\|\dot{x}(\cdot)\|_{L_2(D)} \leq 1$, where $\dot{x}(t) = (\partial x(t)/\partial t_1, \partial x(t)/\partial t_2)$, $t = (t_1, t_2)$.

Consider the problem of optimal recovery of the integral

$$Jx(\cdot) = \iint_D x(t)\, dt_1 dt_2$$

on the class $W_2^1(D)$ based on precisely specified values of the function on the boundary ∂D.

THEOREM 6. *Let $\widehat{x}(\cdot)$ be the solution of the Poisson problem*

$$\Delta x = -1, \quad x\big|_{\partial D} = 0, \qquad (i)$$

where Δ is the Laplace operator. Then:

(a) $E(J, W_2^1(D), F_\partial) = \|\dot{\widehat{x}}(\cdot)\|_{L_2(D)}$;

(b) *the functional*

$$x(\cdot) \to \int_{\partial D} x(t)\frac{\partial \widehat{x}(t)}{\partial t_2}\, dt_1 - x(t)\frac{\partial \widehat{x}(t)}{\partial t_1}\, dt_2$$

is the optimal recovery method.

PROOF. Let $x(\cdot) \in W_2^1(D)$. Set

$$P(t) = -x(t)\frac{\partial \widehat{x}(t)}{\partial t_2}, \quad Q(t) = x(t)\frac{\partial \widehat{x}(t)}{\partial t_1}.$$

Then using Green's formula

$$\int_{\partial D} P(t)\, dt_1 + Q(t)\, dt_2 = \iint_D \left(\frac{\partial Q(t)}{\partial t_1} - \frac{\partial P(t)}{\partial t_2}\right) dt,$$

we obtain

$$\int_{\partial D} -x(t)\frac{\partial \widehat{x}(t)}{\partial t_2}\, dt_1 + x(t)\frac{\partial \widehat{x}(t)}{\partial t_1}\, dt_2 = \iint_D x(t)\Delta\widehat{x}(t)\, dt + \iint_D \dot{x}(t)\dot{\widehat{x}}(t)\, dt.$$

This implies that for all $x(\cdot) \in W_2^1(D)$

$$\iint_D x(t)\,dt = \int_{\partial D} x(t)\frac{\partial \widehat{x}(t)}{\partial t_2}\,dt_1 - x(t)\frac{\partial \widehat{x}(t)}{\partial t_1}\,dt_2 + \iint_D \dot{x}(t)\dot{\widehat{x}}(t)\,dt. \tag{ii}$$

With the aid of the Cauchy–Bunyakovskii inequality we obtain

$$\left| \iint_D x(t)\,dt - \int_{\partial D} x(t)\frac{\partial \widehat{x}(t)}{\partial t_2}\,dt_1 - x(t)\frac{\partial \widehat{x}(t)}{\partial t_1}\,dt_2 \right| \le \|\dot{\widehat{x}}(\cdot)\|_{L_2(D)}. \tag{iii}$$

Hence

$$E(J, W_2^1(D), F_\partial) \le \|\dot{\widehat{x}}(\cdot)\|_{L_2(D)}. \tag{iv}$$

The associated problem to the recovery problem under consideration has the form

$$\iint_D x(t)\,dt \to \max, \qquad x(\cdot)\big|_{\partial D} = 0, \quad x(\cdot) \in W_2^2(D).$$

Using (ii) one can easily obtain a lower bound for its value:

$$\sup_{\substack{x(\cdot)\in W_2^1(D)\\ x(\cdot)|_{\partial D}=0}} \left| \iint_D x(t)\,dt \right| \ge \left| \iint_D \frac{\widehat{x}(t)}{\|\dot{\widehat{x}}(\cdot)\|_{L_2(D)}}\,dt \right| = \|\dot{\widehat{x}}(\cdot)\|_{L_2(D)}.$$

Since the value of the associated problem is no greater than the error of the optimal recovery, this bound and (iv) imply (a), while (b) follows from (iii). □

Consider the particular case where D is the unit disk. One can easily check that the function $\widehat{x}(t) = (1 - |t|^2)/4$ is a solution of the problem (i). In this case we have

$$E(J, W_2^1(D), F_\partial) = \frac{1}{2}\left(\iint_D |t|^2\,dt \right)^{1/2} = \frac{1}{2}\sqrt{\frac{\pi}{2}}.$$

By a passage to polar coordinates the optimal method becomes

$$\iint_D x\,dt \approx \frac{1}{2}\int_0^{2\pi} x(\cos\varphi, \sin\varphi)\,d\varphi.$$

7. *Inequalities for derivatives of smooth functions.* Recall that Kolmogorov-type inequalities for derivatives of smooth functions were considered in the previous section. These are inequalities of the form

$$\|x^{(k)}(\cdot)\|_{L_q(T)} \le K\|x(\cdot)\|^{\alpha}_{L_p(T)}\|x^{(n)}(\cdot)\|^{\beta}_{L_r(T)},$$

where $0 \le k < n$ are integers, $1 \le p, q, r \le \infty$, $\alpha, \beta \ge 0$, $T = \mathbb{R}$ or $\mathbb{R}_+$. They are considered on the space $\mathcal{W}^n_{p,r}(T)$ of functions $x(\cdot) \in L_p(T)$ such that the $(n-1)$th derivative is locally absolutely continuous on T and $x^{(n)}(\cdot) \in L_r(T)$.

As we pointed out, finding the exact constant in such an inequality for $q = \infty$ is equivalent, for example, to the solution of the problem

$$x^{(k)}(0) \to \max, \qquad \|x(\cdot)\|_{L_p(T)} \le \delta, \quad \|x^{(n)}(\cdot)\|_{L_r(T)} \le 1. \qquad (i)$$

This problem may be viewed as the one associated to the optimal recovery problem of the kth derivative of $x(\cdot)$ at zero on the class $W^n_{p,r}(T) = \{x(\cdot) \in \mathcal{W}^n_{p,r}(T) \mid \|x^{(n)}(\cdot)\|_{L_r(T)} \le 1\}$ based on the following information: the function $x(\cdot) \in W^n_{p,r}(T)$ is known to within δ in the metric of $L_p(T)$; i.e., we observe a function $y(\cdot) \in L_p(T)$ which differs from $x(\cdot)$ in this metric by δ. Denote the corresponding information operator by $F_{\delta L_p(T)}$.

For illustration, we will find the optimal recovery methods in some special cases. To this end one could employ the general theorems proved above, but a simpler way will be to use the results obtained in Section 9.

Let, for example, $n = 1$, $k = 0$, $1 < p = r < \infty$, and $T = \mathbb{R}_+$. We will show that the functional

$$y(\cdot) \to \frac{p-1}{\delta} \int_{\mathbb{R}_+} e^{-\frac{p-1}{\delta}t} y(t)\, dt \qquad (ii)$$

yields the optimal method in the corresponding recovery problem. Indeed, let $x(\cdot) \in W^1_{p,p}(\mathbb{R}_+)$ and let $y(\cdot) \in L_p(\mathbb{R}_+)$ be such that $\|x(\cdot) - y(\cdot)\|_{L_p(\mathbb{R}_+)} \le \delta$. In Section 9 (Example 2) the following identity was obtained:

$$x(0) = \frac{p-1}{\delta} \int_{\mathbb{R}_+} e^{-\frac{p-1}{\delta}t} x(t)\, dt - \int_{\mathbb{R}_+} e^{-\frac{p-1}{\delta}t} \dot{x}(t)\, dt, \qquad (iii)$$

which is valid for all functions in $\mathcal{W}^1_{p,p}(\mathbb{R}_+)$. Set $z(\cdot) = x(\cdot) - y(\cdot)$. Then using (iii) we get

$$\begin{aligned} x(0) &- \frac{p-1}{\delta} \int_{\mathbb{R}_+} e^{-\frac{p-1}{\delta}t} y(t)\, dt \\ &= x(0) - \frac{p-1}{\delta} \int_{\mathbb{R}_+} e^{-\frac{p-1}{\delta}t} x(t)\, dt + \frac{p-1}{\delta} \int_{\mathbb{R}_+} e^{-\frac{p-1}{\delta}t} z(t)\, dt \\ &= -\int_{\mathbb{R}_+} e^{-\frac{p-1}{\delta}t} \dot{x}(t)\, dt + \frac{p-1}{\delta} \int_{\mathbb{R}_+} e^{-\frac{p-1}{\delta}t} z(t)\, dt. \end{aligned}$$

Estimating the integrals in the right-hand side by Hölder's inequality, we obtain

$$\left| x(0) - \frac{p-1}{\delta} \int_{\mathbb{R}_+} e^{-\frac{p-1}{\delta}t} y(t)\, dt \right| \le \delta^{1/p'} p^{1/p}. \qquad (iv)$$

The associated problem, i.e., problem (i) for the parameters at hand, was solved in Section 9 (Example 2), and its value was precisely the quantity in the right-hand side of (iv). But we know (see (3)) that the value of the associated problem is always not greater than the optimal recovery error. Together with (iv) this implies that (ii) is the optimal recovery method and $E(x(0), W^1_{p,p}(\mathbb{R}_+), F_{\delta L_p(\mathbb{R}_+)}) = \delta^{1/p'} p^{1/p}$.

Consider one more example where the same approach (based on the results of the previous section) yields the optimal method and the optimal recovery error. The results will be summarized in the following theorem.

THEOREM 7. *Let $n = 1$, $k = 0$, $1 < p = r < \infty$, and $T = \mathbb{R}_+$. Then*

(a) $E(x(0), W^1_{p,p}(\mathbb{R}_+), F_{\delta L_p(\mathbb{R}_+)}) = \delta^{1/p'} p^{1/p}$;

(b) *the linear functional*

$$y(\cdot) \to \frac{p-1}{\delta} \int_{\mathbb{R}_+} e^{-\frac{p-1}{\delta}t} y(t)\, dt$$

yields the optimal recovery method.

Let $n = 2$, $k = 0$, $p = \infty$, $1 < r \le \infty$, and $T = \mathbb{R}_+$. Then

(a) $E(\dot{x}(0), W^2_{\infty,r}(\mathbb{R}_+), F_{\delta L_\infty(\mathbb{R}_+)}) = 2\delta(r'+1)\tau^{-1}$, *where*

$$\tau = \tau(r, \delta) = \left(2r'(r'+1)^{1/r'}\right)^{r'/(r'+1)} \delta^{r'/(r'+1)};$$

(b) *the linear functional*

$$y(\cdot) \to \frac{y(\tau) - y(0)}{\tau}$$

yields the optimal recovery method.

8. *Recovery of the value of a periodic function from its Fourier coefficients.* Let $\mathbb{T}$ be the unit circle (for which we take the interval $[-\pi, \pi]$ identifying its end-points), and let $r \in \mathbb{N}$ and $1 \le p \le \infty$. Denote by $\mathcal{W}^r_p(\mathbb{T})$ the Sobolev space of real-valued functions $x(\cdot)$ on

$\mathbb{T}$ with absolute continuous $(r-1)$th derivative and $x^{(r)}(\cdot) \in L_p(\mathbb{T})$. Any function $x(\cdot) \in \mathcal{W}_p^r(\mathbb{T})$ can be represented as

$$x(\cdot) = \frac{1}{2\pi}\int_{\mathbb{T}} x(t)\,dt + \int_{\mathbb{T}} B_r(\cdot - t)x^{(r)}(t)\,dt, \qquad \int_{\mathbb{T}} x^{(r)}(t)\,dt = 0, \quad (i)$$

where $B_r(t) = \frac{1}{\pi}\sum_{k=1}^{\infty} k^{-r}\cos(kt - \pi r/2)$ is the Bernoulli kernel. This is an even function for even r and an odd function for odd r.

The set $W_p^r(\mathbb{T}) = \{x(\cdot) \in \mathcal{W}_p^r(\mathbb{T}) \mid \|x^{(r)}(\cdot)\|_{L_p(\mathbb{T})} \le 1\}$ is called the Sobolev class of functions.

We are going to consider the following problem: *recover the value of the function $x(\cdot)$ at a point $\tau \in \mathbb{T}$ on the class $W_p^r(\mathbb{T})$ based on the information that the first $2n-1$ Fourier coefficients of $x(\cdot)$ are known.*

In accordance with our general notations we have here $X = \mathcal{W}_p^r(\mathbb{T})$, $C = W_p^r(\mathbb{T})$, $Y = \mathbb{R}^{2n-1}$, $Fx(\cdot) = F_{2n-1}x(\cdot) = (a_0, a_1, \dots, a_{n-1}, b_1, \dots, b_{n-1})$, where $a_0 = \frac{1}{\pi}\int_{\mathbb{T}} x(t)\,dt$,

$$a_k = \frac{1}{\pi}\int_{\mathbb{T}} x(t)\cos kt\,dt, \quad b_k = \frac{1}{\pi}\int_{\mathbb{T}} x(t)\sin kt\,dt, \quad k = 1, \dots, n-1,$$

and $\langle x', x(\cdot)\rangle = x(\tau)$.

THEOREM 8. *Let $r \in \mathbb{N}$, $\tau \in \mathbb{T}$. If $1 < p \le \infty$ and $\widehat{\xi}_{rp'n}(\cdot) = \alpha_0 + \sum_{k=1}^{n-1}(\alpha_{knrp'}\cos k\cdot + \beta_{knrp'}\sin k\cdot)$ is the best approximation polynomial of the function $B_r(\cdot)$ by the subspace $\mathcal{T}_{n-1}$ in the metric of $L_{p'}(\mathbb{T})$, then*

$$x(\tau) \approx \frac{a_0}{2} + \sum_{k=1}^{n-1} c_{knrp'}k^r(a_k\cos k\tau + b_k\sin k\tau),$$

where $c_{knrp'} = (-1)^{r/2}\alpha_k$ if r is even and $c_{knrp'} = (-1)^{(r-1)/2}\beta_k$ if r is odd.

Moreover,

$$E(x(\tau), W_p^r(\mathbb{T}), F_{2n-1}) = d(B_r(\cdot), \mathcal{T}_{n-1}, L_{p'}(\mathbb{T})) \qquad (*)$$

and

$$d(W_\infty^r(\mathbb{T}), \mathcal{T}_{n-1}, C(\mathbb{T})) = E(x(\tau), W_\infty^r(\mathbb{T}), F_{2n-1}) = \frac{K_r}{n^r}, \qquad (**)$$

where $K_r = \frac{4}{\pi}\sum_{j=0}^{\infty}\frac{(-1)^{j(r+1)}}{(2j+1)^{r+1}}$.

PROOF. The associated problem to the recovery problem under consideration has the form

$$x(\tau) \to \max, \quad a_0 = a_1 = \cdots = a_{n-1} = b_1 = \cdots = b_{n-1} = 0, \quad x(\cdot) \in W_p^r(\mathbb{T}). \tag{ii}$$

Along with (ii) consider the problem

$$x(0) \to \max, \quad a_1 = \cdots = a_{n-1} = b_1 = \cdots = b_{n-1} = 0, \quad x(\cdot) \in \mathring{W}_p^r(\mathbb{T}), \tag{iii}$$

where $\mathring{W}_p^r(\mathbb{T}) = \{x(\cdot) \in W_p^r(\mathbb{T}) \mid \int_{\mathbb{T}} x(t)\,dt = 0\}$.

Henceforth we assume for definiteness that r is even. Then denoting $u(\cdot) = x^{(r)}(\cdot)$ and integrating by parts (using (i)), the problem (iii) becomes

$$\int_{\mathbb{T}} B_r(t)u(t)\,dt \to \max, \qquad \int_{\mathbb{T}} u(t)\,dt = 0,$$
$$\int_{\mathbb{T}} u(t)\cos kt\,dt = 0, \quad \int_{\mathbb{T}} u(t)\sin kt\,dt = 0, \quad 1 \le k \le n-1, \tag{iv}$$
$$\|u(\cdot)\|_{L_p(\mathbb{T})} \le 1.$$

This problem has the form (P_I^*) and therefore is dual to the distance problem

$$\|B_r(\cdot) - \xi(\cdot)\|_{L_{p'}(\mathbb{T})} \to \min, \quad \xi(\cdot) \in \mathcal{T}_{n-1}, \tag{v}$$

where $\mathcal{T}_{n-1}$ is the space of trigonometric polynomials of degree not exceeding $n-1$.

We proceed in accordance with the general scheme. Let $\widehat{\xi}(\cdot) \in \mathcal{T}_{n-1}$ and let $\widehat{u}(\cdot)$ be an admissible function in (iv). Denote $\widehat{z}(\cdot) = B_r(\cdot) - \widehat{\xi}(\cdot)$ and assume that at $\widehat{\xi}(\cdot)$ and $\widehat{u}(\cdot)$ the equality between the values of problems (iv) and (v) is attained, which in view of $\widehat{u}(\cdot) \in \mathcal{T}_{n-1}^{\perp}$ may be written as

$$\int_{\mathbb{T}} \widehat{z}(t)\widehat{u}(t)\,dt = \|\widehat{z}(\cdot)\|_{L_{p'}(\mathbb{T})}. \tag{vi}$$

This implies that function $\widehat{u}(\cdot)$ is proportional to $|\widehat{z}(\cdot)|^{p'-1}\operatorname{sign}\widehat{z}(\cdot)$; hence the orthogonality condition of $\widehat{u}(\cdot)$ to the subspace $\mathcal{T}_{n-1}$ has the form

$$\begin{aligned}&\int_{\mathbb{T}} |\widehat{z}(t)|^{p'-1} \operatorname{sign} \widehat{z}(t)\, dt = 0,\\ &\int_{\mathbb{T}} |\widehat{z}(t)|^{p'-1} \operatorname{sign} \widehat{z}(t) \begin{matrix}\cos kt\\ \sin kt\end{matrix}\, dt = 0, \quad 1 \le k \le n-1.\end{aligned} \tag{vii}$$

Note that (v) is a smooth minimization problem (since $\widehat{z}(\cdot) \neq 0$, for $p' = 1$ the problem is also smooth) of a function of $2n-1$ variables (coefficients of $\xi(\cdot)$); hence the partial derivatives of the function to be minimized must vanish at the point of minimum, which is expressed by relation (vii).

Set $\widehat{u}(\cdot) = \|\widehat{z}(\cdot)\|^{1-p'}_{L_{p'}(\mathbb{T})} |\widehat{z}(\cdot)|^{p'-1} \operatorname{sign} \widehat{z}(\cdot)$. This is a function admissible in the problem (iv), and obviously the equality (vi) holds. Therefore $\widehat{u}(\cdot)$ is a solution to the problem (iv). This problem is associated with the recovery problem of the functional $u(\cdot) \to \int_{\mathbb{T}} B_r(t)u(t)dt$ on the class $BL_p(\mathbb{T})$ with information operator

$$Fu(\cdot) = \left(\int_{\mathbb{T}} u(t)dt, \int_{\mathbb{T}} u(t)\cos ktdt, \int_{\mathbb{T}} u(t)\sin ktdt,\right.$$
$$\left. k = 1, \dots, n-1\right).$$

Then by the general theorem

$$\int_{\mathbb{T}} B_r(t)u(t)dt \approx \int_{\mathbb{T}} (\alpha_0 + \sum_{k=1}^{n-1} \alpha_k \cos kt)u(t)dt,$$

where, recall, the α_k are the coefficients of the best approximation polynomial $B_r(\cdot)$. It is not hard to see that $\beta_k = 0$, $k = 1, \dots, n-1$, since $B_r(\cdot)$ is an even function. Now substituting $x^{(r)}(\cdot)$ for $u(\cdot)$ and integrating by parts, we obtain that in the recovery problem corresponding to (iii), $x(0) \approx (-1)^{r/2}k^r \sum_{k=1}^{n-1} \alpha_k a_k$. Then we have in the initial recovery problem for $\tau = 0$ that

$$x(0) \approx a_0 + (-1)^{r/2}k^r \sum_{k=1}^{n-1} \alpha_k a_k.$$

Indeed, recall that $x(0) = a_0/2 + \int_{\mathbb{T}} B_r(t)x^{(r)}(t)dt$ and denote $a = (a_0, a_1, \dots, a_{n-1}, b_1, \dots, b_{n-1})$ and $\overline{a} = (a_1, \dots, a_{n-1}, b_1, \dots, b_{n-1}))$.

Then for an arbitrary function φ on $\mathbb{R}^{2n-1}$ we have

$$\sup_{x(\cdot)\in W_p^r(\mathbb{T})} |x(0)-\varphi(a)| \geq \sup_{x(\cdot)\in \mathring{W}_p^r(\mathbb{T})} \left| \int_{\mathbb{T}} B_r(t)x^{(r)}(t)dt - \varphi(0,\overline{a})\right|$$

$$\geq E(x(0), \mathring{W}_p^r(\mathbb{T}), F_0)$$

$$= \sup_{x(\cdot)\in \mathring{W}_p^r(\mathbb{T})} \left| \int_{\mathbb{T}} B_r(t)x^{(r)}(t)dt - (-1)^{r/2}k^r \sum_{k=1}^{n-1} \alpha_k a_k \right|$$

$$= \sup_{x(\cdot)\in W_p^r(\mathbb{T})} \left| x(0) - \left(\frac{a_0}{2} + (-1)^{r/2}k^r \sum_{k=1}^{n-1} \alpha_k a_k\right)\right|,$$

which proves the required assertion.

The case of odd r is treated similarly. Due to translation invariance of the class $W_p^r(\mathbb{T})$ the recovery method of $x(\cdot)$ at zero can be easily carried over to any point $\tau \in \mathbb{T}$.

Now we prove the equality $(**)$. The Bernoulli kernel possesses the Favard property: for even (odd) r the difference between $B_r(\cdot)$ and the even (odd) polynomial interpolating $B_r(\cdot)$ at the roots of $\cos n\,\cdot$ $(\sin n\,\cdot)$ has no other roots besides those of $\cos n\,\cdot$ $(\sin n\,\cdot)$. This is a well-known fact (see, e.g., Akhiezer (1965)) which follows directly from Rolle's theorem. Again let r be even and let $\widehat{q}(\cdot)$ be the corresponding interpolation polynomial. Then $\widehat{u}(\cdot) = \pm\operatorname{sign}\cos n\,\cdot$. It is not hard to see that $\widehat{u}(\cdot) \in \mathcal{T}_{n-1}^{\perp}$; hence $\widehat{q}(\cdot)$ is the best approximation polynomial for $B_r(\cdot)$ in the metric of $L_1(\mathbb{T})$. But, as was pointed out before, such polynomial is unique, so that $\widehat{q}(\cdot) = \widehat{p}(\cdot)$. Thus

$$d\left(B_r(\cdot), \mathcal{T}_{n-1}, L_1(\mathbb{T})\right) = \int_{\mathbb{T}} |B_r(t) - \widehat{p}(t)|\,dt$$

$$= \pm \int_{\mathbb{T}} (B_r(t) - \widehat{p}(t))\operatorname{sign}\cos nt dt$$

$$= \left| \int_{\mathbb{T}} B_r(t)\operatorname{sign}\cos nt dt\right| = \frac{\pi K_r}{n^r}.$$

The last equality is obtained by a direct calculation. Combined with $(*)$ it implies the second equality in $(**)$. The proof of $(**)$ is completed by taking the supremum in $\tau \in \mathbb{T}$. □

9. *Recovery of the derivative of a function from its approximate Fourier coefficients.* In this part we deal with recovery of a linear operator rather than a functional as before. But in this situation the methods of convex analysis play a key role as well. We did not state in this section any general results on recovery of operators, so that we will solve this problem directly. For further generalization and development of the problem under consideration see Magaril-Il'yaev and Osipenko (2002), (2003).

Let $W_2^2(\mathbb{T})$ be a Sobolev class of 2π-periodic functions (generally, complex-valued) defined as in the previous example. On this class we want to recover the first derivative of a function $x(\cdot)$ in the metric of $L_2(\mathbb{T})$ based on a finite collection of its Fourier coefficients

$$x_j = \frac{1}{2\pi}\int_{\mathbb{T}} x(t)e^{-ijt}\,dt$$

observed with an error. More precisely, we will assume that for each function $x(\cdot) \in W_2^2(\mathbb{T})$ we know numbers y_j, $|j| \le N$, such that $|x_j - y_j| \le \delta$, $|j| \le N$, $\delta > 0$; i.e., the information operator is the multi-valued mapping F_δ^N which associates with each function $x(\cdot) \in W_2^2(\mathbb{T})$ the set $F_\delta^N x(\cdot) = \{y = \{y_j\}_{|j|\le N} \mid |x_j - y_j| \le \delta, |j| \le N, \delta > 0\}$. The problem is to determine

$$E(\dot{x}(\cdot), W_2^2(\mathbb{T}), F_\delta^N) = \inf_{\varphi\colon\, \mathbb{C}^{2N+1}\to L_2(\mathbb{T})} \sup_{\substack{x(\cdot)\in W_2^2(\mathbb{T})\\ y\in F_\delta^N x(\cdot)}} \|\dot{x}(\cdot) - \varphi(y)(\cdot)\|_{L_2(\mathbb{T})}$$

and the corresponding optimal method.

THEOREM 9. *Let*

$$p_0 = \max\Big\{ p \in \mathbb{Z}_+ \;\Big|\; \delta^2 \sum_{|j|\le p} j^4 < 1,\ 0 \le p \le N \Big\}.$$

Then:

(*a*)

$$E(\dot{x}(\cdot), W_2^2(\mathbb{T}), F_\delta^N) = \Big(1 + \delta^2 \sum_{|j|\le p_0} \big(j^2(p_0+1)^2 - j^4\big)\Big)^{1/2} / (p_0+1);$$

(b) *the operator that with a set* $y = \{y_j\}_{|j|\le N}$ *associates the function*

$$\widehat{\varphi}(y)(t) = \sum_{|j|\le p_0} ij\left(1 - \left(\frac{j}{p_0+1}\right)^2\right) y_j e^{ijt}$$

yields the optimal recovery method.

PROOF. Similarly to (3) one can easily prove the lower bound

$$E(\dot{x}(\cdot), W_2^2(\mathbb{T}), F_\delta^N) \ge \sup_{\substack{x(\cdot)\in W_2^2(\mathbb{T}) \\ |x_j|\le\delta,\ |j|\le N}} \|\dot{x}(\cdot)\|_{L_2(\mathbb{T})}. \qquad (i)$$

By Parseval's identity the right-hand side of (i) is rewritten (in terms of squared norm) as

$$\sum_{j\in\mathbb{Z}} j^2 u_j \to \max, \qquad \sum_{j\in\mathbb{Z}} j^4 u_j \le 1, \quad 0 \le u_j \le \delta^2,\ |j| \le N, \qquad (ii)$$

where $u_j = |x_j|^2$, $j \in \mathbb{Z}$.

Now, (ii) is a convex programming problem. According to the remark after the proof of the Karush–Kuhn–Tucker theorem (Section 5), to solve this problem it suffices to find $\lambda \ge 0$, $\lambda_j \ge 0$, $|j| \le N$, and an admissible sequence $\{\widehat{u}_j\}_{j\in\mathbb{Z}}$ such that for all $u_j \ge 0$, $j \in \mathbb{Z}$,

(a) $\sum_{j\in\mathbb{Z}}(-j^2 + \lambda j^4 + \lambda_j\chi_j)u_j \ge \sum_{j\in\mathbb{Z}}(-j^2 + \lambda j^4 + \lambda_j\chi_j)\widehat{u}_j$,

(b) $\lambda\Big(\sum_{j\in\mathbb{Z}} j^4\widehat{u}_j - 1\Big) = 0, \quad \lambda_j(\widehat{u}_j - \delta_j^2) = 0,\ |j| \le N,$

where $\chi_j = 1$ for $|j| \le N$ and $\chi_j = 0$ otherwise. Let p_0 be as in the theorem. Set $\lambda = (p_0+1)^{-2}$, $\lambda_j = j^2 - (p_0+1)^{-2}j^4$ for $|j| \le p_0$ and $\lambda_j = 0$ for $p_0 + 1 \le |j| \le N$. Define the sequence $\widehat{u} = \{\widehat{u}_j\}_{j\in\mathbb{Z}}$ as follows: $\widehat{u}_j = \delta^2$ for $|j| \le p_0$,

$$\widehat{u}_j = \frac{1 - \delta^2\sum_{|k|\le p_0} k^4}{2(p_0+1)^4} \qquad \text{for} \quad |j| = p_0 + 1,$$

and $\widehat{u}_j = 0$ for $|j| > p_0+1$. It can be easily verified that this sequence is admissible in the problem (ii) and fulfills conditions (a) and (b). Therefore $\widehat{u}$ is the solution to the problem (ii). Putting $\widehat{u}$ into the functional to be maximized and extracting the square root, we obtain in view of (i) that

$$E(\dot{x}(\cdot), W_2^2(\mathbb{T}), F_\delta^N) \ge \frac{\left(1 + \delta^2\sum_{|j|\le p_0}\left(j^2(p_0+1)^2 - j^4\right)\right)^{1/2}}{p_0+1}. \qquad (iii)$$

Again, using sufficient conditions in the Karush–Kuhn–Tucker theorem, one can see that $\widehat{u}$ at the same time solves the problem

$$\sum_{j\in\mathbb{Z}} j^2 u_j \to \max, \quad \lambda\sum_{j\in\mathbb{Z}} j^4 u_j + \sum_{|j|\le N} \lambda_j u_j \le \lambda + \delta^2 \sum_{j\in\mathbb{Z}} \lambda_j, \ u_j \ge 0. \tag{iv}$$

Therefore the problems (ii) and (iv) have the same values.

Set $\widehat{x}_j = y_0$ for $j = 0$,

$$\widehat{x}_j = \frac{\lambda_j}{\lambda j^4 + \lambda_j} y_j \qquad \text{for} \quad 1 \le |j| \le p_0,$$

and $\widehat{x}_j = 0$ for $|j| > p_0$.

It can be directly verified that for all $x(\cdot) \in W_2^2(\mathbb{T})$

$$\lambda\sum_{j\in\mathbb{Z}} j^4 |x_j - \widehat{x}_j|^2 + \sum_{|j|\le N} \lambda_j |x_j - \widehat{x}_j|^2 + \lambda\sum_{j\in\mathbb{Z}} j^4 |\widehat{x}_j|^2 + \sum_{|j|\le N} \lambda_j |\widehat{x}_j - y_j|^2$$
$$= \lambda\sum_{j\in\mathbb{Z}} j^4 |x_j|^2 + \sum_{|j|\le N} \lambda_j |x_j - y_j|^2.$$

If $x(\cdot) \in W_2^2(\mathbb{T})$ and $|x_j - y_j| \le \delta$, then setting $v_j = |x_j - \widehat{x}_j|^2$ we will have

$$\lambda\sum_{j\in\mathbb{Z}} j^4 v_j + \sum_{|j|\le N} \lambda_j v_j \le \lambda\sum_{j\in\mathbb{Z}} j^4 |x_j|^2 + \sum_{|j|\le N} \lambda_j |x_j - y_j|^2$$
$$\le \lambda + \delta^2 \sum_{|j|\le N} \lambda_j.$$

Hence

$$\left\| \dot{x}(t) - \sum_{|j|\le N} ij\widehat{x}_j e^{ijt} \right\|_{L_2(\mathbb{T})}^2 = \sum_{j\in\mathbb{Z}} j^2 v_j$$
$$\le \sup\left\{ \sum_{j\in\mathbb{Z}} j^2 u_j \,\middle|\, \lambda\sum_{j\in\mathbb{Z}} j^4 u_j + \sum_{|j|\le N} \lambda_j u_j \le \lambda + \delta^2 \sum_{|j|\le N} \lambda_j, \ u_j \ge 0 \right\}.$$

Since the value of the extremal problem in the right-hand side, which is in fact problem (iv), is equal to the value of problem (ii), we obtain an upper bound for the optimal recovery error which coincides with the lower bound (iii), and the operator which associates the function

$$\widehat{\varphi}(y)(t) = \sum_{|j|\leq N} ij\widehat{x}_j e^{ijt} = \sum_{|j|\leq p_0} ij\left(1 - \left(\frac{j}{p_0+1}\right)^2\right) y_j e^{ijt}$$

with the set $y = \{y_j\}_{|j|\leq N}$ yields the optimal recovery method. □

Note that if $p_0 < N$, then the optimal recovery error does not reduce anymore with increasing the number of the Fourier coefficients observed with the same error. Thus for a fixed δ the set of $2N(\delta)$ Fourier coefficients (the zero coefficient is not used in the optimal method), where

$$N(\delta) = \max\Big\{ N \in \mathbb{Z}_+ \;\Big|\; \delta^2 \sum_{|j|\leq N} j^4 < 1 \Big\},$$

allows for recovery of the derivative of a function in $L_2(\mathbb{T})$ with maximal accuracy.

CHAPTER 3

Appendix

11. Basic theorems of convex analysis

In this section we collected some theorems which are basic for many facts and phenomena of convex analysis.

Let X be a topological vector space and X^* its dual. A set $H \subset X^*$ is *equicontinuous* if for any $\varepsilon > 0$ there is a neighborhood of zero U in X such that $|\langle x^*, x\rangle| < \varepsilon$ for all $x \in U$ and $x^* \in H$. Recall that the weak topologies $\sigma(X, X^*)$ and $\sigma(X^*, X)$ in case X is a locally convex space were defined in Section 6.

THEOREM (Banach–Alaoglu). *Let X be a locally convex space and X^* its dual. Then any equicontinuous subset of X^* is relatively compact in the topology $\sigma(X^*, X)$.*

PROOF. The proof relies on Tychonoff's theorem. Recall that if $\mathcal{A}$ is some set and to each $\alpha \in \mathcal{A}$ there corresponds a nonempty set X_α, then the set of all functions x on $\mathcal{A}$ such that $x(\alpha) \in X_\alpha$ is called the (Cartesian) product of the sets X_α, denoted by $\prod_{\alpha\in\mathcal{A}} X_\alpha$. If the X_α are topological spaces, then the *Tychonoff topology* in the product space is defined as follows: the base of open sets consists of the sets of the form $\prod_{\alpha\in\mathcal{A}} G_\alpha$, where G_α is an open subset of X_α with $G_\alpha = X_\alpha$ except for finitely many α's. Tychonoff's theorem says that *the product of compact topological spaces is compact.*

Now we proceed to the proof of the theorem. Let, for definiteness, X be a real space and let H be an equicontinuous subset of X. It is easy to see that the topology $\sigma(X^*, X)$ is induced by the topology of the product $\mathbb{R}^X = \prod_{x\in X} \mathbb{R}_x$, $\mathbb{R}_x = \mathbb{R}$. Since H is equicontinuous, there is a neighborhood of zero $U \subset X$ such that $|\langle x^*, x\rangle| \le 1$ for all $x \in U$ and $x^* \in H$. Let $x_0 \in X$. Since the operation of multiplication by numbers is continuous, there is $\lambda > 0$ such that $\lambda x_0 \in U$; i.e., $|\langle x^*, x_0\rangle| \le \lambda^{-1}$ for all $x^* \in H$. Thus the set $\{\langle x^*, x\rangle \mid x^* \in H\}$ is bounded in $\mathbb{R}$ and hence relatively compact. By Tychonoff's theorem

$\operatorname{cl} H$ is compact in $\mathbb{R}^X$; hence H is relatively compact in $\mathbb{R}^X$. But it is not hard to check that $\operatorname{cl} H \subset X^*$; therefore H is relatively compact in X^*. □

Let us point out a frequently used corollary to this theorem: *a bounded subset of the space X^* dual to a normed space X is relatively compact in topology $\sigma(X^*, X)$.*

Let A be a subset of a vector space X. A nonempty set $M \subset A$ is an *extremal subset* of A if neither point of M is an interior point of an interval with end-points in A but not in M. In other words, $x, y \in A$ and $(1-\alpha)x + \alpha y \in M$ for some $\alpha \in (0,1)$ entail $x, y \in M$.

A single-point extremal set is an *extreme point.*

THEOREM (Minkowski–M. Krein–D. Milman). *Any nonempty convex compact set in a locally convex space coincides with the closure of the convex hull of its extreme points.*

PROOF. Let K be a nonempty compact subset of a locally convex space X. We will show first that K contains at least one extreme point. The set K is an extremal subset of itself. Denote by $\mathcal{M}$ the class of all closed extremal subsets of K ordered by inclusion. If $\mathcal{M}_1$ is a linearly ordered subset of $\mathcal{M}$, then the set $\bigcap_{M \in \mathcal{M}_1} M$ is nonempty (being the intersection of a centered family of closed sets). It is easily verified that this is an extremal subset of K, and it is obviously a lower bound for $\mathcal{M}_1$. By Zorn's lemma, there exists the minimal element M_0 of $\mathcal{M}$. Let us show that M_0 is an extreme point of K. Indeed, assume that M_0 contains two different points x_1 and x_2. By the second separation theorem (see Section 1.3) there is a linear continuous functional x^* on X such that $\langle x^*, x_1 \rangle \neq \langle x^*, x_2 \rangle$. This implies that $M_1 = \{x \in M_0 \mid \langle x^*, x \rangle = \min_{y \in M_0} \langle x^*, y \rangle\}$ is a proper closed subset of M_0. On the other hand, if $(1-\alpha)y_1 + \alpha y_2 \in M_1$ for some points y_1 and y_2 in K and $\alpha \in (0,1)$, then, since M_0 is an extremal subset, the points y_1 and y_2 belong to M_0. By definition of M_1 we immediately get that $y_1, y_2 \in M_1$, so that M_1 is an extremal closed proper subset of M_0, which contradicts minimality of M_0. Thus M_0 consists of a single point. Note that we have actually proved that any extremal closed set K has an extreme point (without using convexity of K).

Denote by E the closure of the convex hull of the set of extreme points of K. Clearly, $E \subset K$. Assume that $E \neq K$ and let $x_0 \in K \setminus E$. By the second separation theorem there is a functional $x_0^* \in X^*$ such that $\sup_{x \in E} \langle x_0^*, x \rangle < \langle x_0^*, x_0 \rangle$. Let $A_0 = \{x \in K \mid \langle x_0^*, x \rangle =$

$\sup_{y \in K}\langle x_0^*, y\rangle\}$. Clearly, this set is disjoint with E and it is easily verified that it is an extremal subset of K. As was proved, A_0 contains an extreme point of K, in contradiction with $A_0 \cap E = \emptyset$. □

Let Σ be a σ-algebra of sets. A real-valued (signed) measure μ on Σ is *nonatomic* if any set $E \in \Sigma$ with $|\mu|(E) > 0$ (where $|\mu|$ is the total variation of μ) contains a subset $A \in \Sigma$ such that $0 < |\mu|(A) < |\mu|(E)$.

THEOREM (A. A. Lyapunov). *Let $\mu_1, \dots, \mu_n$, $n \in \mathbb{N}$, be finite real-valued nonatomic measures on a σ-algebra Σ. Then the image of the mapping $m\colon \Sigma \to \mathbb{R}^n$, $m(E) = (\mu_1(E), \dots, \mu_n(E))$, $E \in \Sigma$, is a convex compact set in $\mathbb{R}^n$.*

PROOF. Let $\mu = |\mu_1| + \dots + |\mu_n|$. It is clear that all the measures μ_i, $i = 1, \dots, n$, are absolutely continuous with respect to μ; hence by the Radon–Nikodim theorem there are functions $\lambda_i(\cdot) \in L_1(T, \mu)$ such that $d\mu_i(t) = \lambda_i(t)d\mu(t)$, $1 \le i \le n$. Consider the set $K = \{x(\cdot) \in L_\infty(T, \mu) \mid 0 \le x(t) \le 1$ for almost all $t \in T\}$. Clearly, $K = \frac{1}{2} + \frac{1}{2}B$, where B is the unit ball in $L_\infty(T, \mu)$; hence due to continuity of the operations and the Bahach–Alaoglu theorem the set K is compact in topology $\sigma\left(L_\infty(T, \mu), L_1(T, \mu)\right)$.

Consider the operator $\Lambda\colon L_\infty(T, \mu) \to \mathbb{R}^n$,

$$\begin{aligned}\Lambda x(\cdot) &= \left(\int_T x(t)d\mu_1(t), \dots, \int_T x(t)d\mu_n(t)\right)\\ &= \left(\int_T x(t)\lambda_1(t)d\mu(t), \dots, \int_T x(t)\lambda_n(t)d\mu(t)\right).\end{aligned}$$

Clearly, this operator is linear and continuous (in topology $\sigma(L_\infty(T, \mu), L_1(T, \mu))$); hence $\Lambda(K)$ is a convex compact set in $\mathbb{R}^n$. Let us show that $m(\Sigma) = \Lambda(K)$. Since the characteristic function of any set $E \in \Sigma$ belongs to K, we have $m(\Sigma) \subset \Lambda(K)$. Let us prove the opposite inclusion. Let $\xi = (\xi_1, \dots, \xi_n) \in \mathbb{R}^n$. Denote $K_\xi = \{x(\cdot) \in K \mid \Lambda x(\cdot) = \xi\}$. Being an inverse image of a convex closed set under a linear continuous mapping, K_ξ is convex and closed, and since $K_\xi \subset K$, it is a convex compact set in topology $\sigma\left(L_\infty(T, \mu), L_1(T, \mu)\right)$. By the Minkowski–Krein–Milman theorem the set K_ξ contains an extreme point $x_0(\cdot)$. We will show that $x_0(\cdot)$ is necessarily a characteristic function of some set $E_0 \in \Sigma$. Then the theorem will follow because $\xi_i = \int_{E_0} d\mu_i(t) = \mu_i(E_0)$, $i = 1, \dots, n$; i.e., $\Lambda(K) \subset m(\Sigma)$, so that $m(\Sigma) = \Lambda(K)$.

Assume that $x_0(\cdot)$ is not a characteristic function. Then for some $0 < \varepsilon < 1$ the set $\Delta = \{t \in T \mid \varepsilon \leq x_0(t) \leq 1 - \varepsilon\}$ is of positive measure. In the space $L_1(\Delta, \mu)$ (which is infinite-dimensional since μ is nonatomic) consider the subspace M generated by restrictions of the functions $\lambda_i(\cdot)$, $i = 1, \dots, n$, to Δ. The subspace M is finite-dimensional and hence closed. The second separation theorem implies that the annihilator of M is nontrivial; i.e., there exists a function $y(\cdot) \in L_\infty(\Delta, \mu)$ such that $\|y(\cdot)\|_{L_\infty(\Delta,\mu)} = 1$ and $\int_\Delta \lambda_i(t) y(t) d\mu(t) = 0$, $i = 1, \dots, n$. Denote by $y_0(\cdot)$ the function on T equal to $y(\cdot)$ on Δ and to zero outside Δ. Then, obviously, $0 \leq x_0(t) \pm \varepsilon y_0(t) \leq 1$ for almost all $t \in T$ and $\Lambda(x_0(\cdot) \pm \varepsilon y_0(\cdot)) = \xi$. This means that $x_0(\cdot)$ is not an extreme point of K_ξ, contrary to the assumption. □

CLEAN UP THEOREM (Levin (1985)). *Let X be a finite-dimensional space, let T be a compact set with* $\operatorname{card} T \geq \dim X + 1$, *and let* $F \colon T \times X \to \mathbb{R}$ *be a function such that*

(a) *the mapping $F(t, \cdot) \colon X \to \mathbb{R}$ is convex for any $t \in T$,*

(b) *the mapping $F(\cdot, x) \colon T \to \mathbb{R}$ is upper semicontinuous for any $x \in X$,*

(c) $M = \inf_{x \in X} \max_{t \in T} F(t, x) > -\infty$.

Then there exist $r \in \mathbb{N}$, $r \leq \dim X + 1$, and points $\tau_1, \dots, \tau_r \in T$ such that $M = \inf_{x \in X} \max_{1 \leq i \leq r} F(\tau_i, x)$.

PROOF. It is clear that $M < +\infty$. With each collection $\tau = (\tau_1, \dots, \tau_{n+1}) \in T^{n+1}$ we associate the number $m(\tau) = \inf_{x \in X} \max_{1 \leq i \leq n+1} F(\tau_i, x)$, and let $m = \sup_{\tau \in T^{n+1}} m(\tau)$. Obviously, $m \leq M$. Let us prove the reverse inequality. Take an $\varepsilon > 0$. For any $t \in T$ consider the set $A(t) = \{x \in X \mid F(t, x) \leq m + \varepsilon\}$. It is easily seen that these are nonempty convex closed subsets of X. We will construct one more set, $\widehat{A}$. Let $\xi = (\xi_1, \dots, \xi_n) \in T^n$. Obviously, there exists an $x(\xi) \in X$ such that $F(\xi_i, x(\xi)) < m + \varepsilon$, $i = 1, \dots, n$. Set $V(\xi) = \{\xi' = (\xi'_1, \dots, \xi'_n) \in T^n \mid F(\xi'_i, x(\xi)) < m + \varepsilon,\ i = 1, \dots, n\}$. Since the function $F(\,\cdot\,, x(\xi))$ is upper semicontinuous, $V(\xi)$ is an open subset of T^n (containing ξ). Now we select a finite subcovering from the open covering $\{V(\xi)\}_{\xi \in T^n}$ of the compact set T^n; let it correspond to the points $\xi^1, \dots, \xi^s$. Consider the set $\widehat{A} = \operatorname{co}\{x(\xi^1), \dots, x(\xi^s)\}$. It is clear that $\widehat{A}$ is a convex compact set.

We will show that any $n + 1$ sets in the family $\{A(t)\}_{t \in T} \cup \widehat{A}$ have a nonempty intersection. Indeed, this is obvious if $\widehat{A}$ is not among these sets. Now consider sets $A(\eta_1), \dots, A(\eta_n), \widehat{A}$, where $\eta = (\eta_1, \dots, \eta_n) \in T^n$. Since $\eta \in V(\xi^i)$ for some $1 \leq i \leq s$, this means that $F(\eta_j, x(\xi^i)) < m + \varepsilon$, $j = 1, \dots, n$, i.e., $x(\xi^i) \in \bigcap_{j=1}^n A(\eta_j)$. But $x(\xi^i) \in \widehat{A}$, so that the family $A(\eta_1), \dots, A(\eta_n), \widehat{A}$ has a nonempty intersection. Then by Helly's theorem $\widehat{A} \cap \bigcap_{t \in T} A(t) \neq \emptyset$. This implies that $M \le m + \varepsilon$ and hence $M = m$.

The function $\tau = (\tau_1, \dots, \tau_{n+1}) \to \max_{1 \le i \le n+1} F(\tau_i, x)$ is upper semicontinuous on T^{n+1} for any $x \in X$. Then $\tau \to m(\tau) = \inf_{x \in X} \max_{1 \le i \le n+1} F(\tau_i, x)$ is also an upper semicontinuous function on T^{n+1}; hence it attains its maximum at some point $\widehat{\tau} = (\widehat{\tau}_1, \dots, \widehat{\tau}_{n+1})$. Taking only different components of $\widehat{\tau}$ we obtain the theorem. □

12. Supplementary topics of convex analysis

In this section we present some fragments of modern theories related in one way or another to the problems of convex analysis discussed in this book.

12.1. On convex analysis of integral functionals.

Many formulas of subdifferential calculus stated in this book (the subdifferential of a sum of finitely many functions, of the maximum of a family of functions, etc.) allow for far-reaching generalization and extension. Here we will state some of them.

Let X be a locally convex space, (T, Σ, μ) a measurable space, and $f_t \colon X \to \mathbb{R} \cup \{+\infty\}$, $t \in T$, a family of convex functions. Further, let V be an open convex subset of X which belongs to all of $\operatorname{dom} f_t$, $t \in T$. Set $\varphi(x) = \int_T f_t(x) d\mu(t)$, $x \in V$. This is a convex function. The following theorem is an analog of the Moreau–Rockafellar theorem for integral functionals.

THEOREM 1. *Let the space X be separable, and let the functions f_t and φ be continuous at $\widehat{x} \in V$. Then*

$$\partial \varphi(\widehat{x}) = \int_T \partial f_t(\widehat{x}) d\mu(t).$$

Now let T be a compact set and $f_t \colon X \to \mathbb{R} \cup \{+\infty\}$, $t \in T$, a family of convex functions with the same set $D = \operatorname{dom} f_t$ for all $t \in T$. Further, assume that for any $x \in D$ the function $t \to f_t$ is lower semicontinuous. Consider the convex function $\varphi(x) = \max_{t \in T} f_t(x)$. Let $\widehat{x} \in D$ and $T(\widehat{x}) = \{t \in T \mid f_t(\widehat{x}) = \varphi(\widehat{x})\}$. Denote by $W(\widehat{x})$ a family of probability measures on T with support in $T(\widehat{x})$. The following is an analog of the Dubovitskii–Milyutin theorem (see Levin (1985)).

THEOREM 2.

$$\partial\varphi(\widehat{x}) = \bigcup_{\mu\in W(\widehat{x})} \partial\left(\int_T f_t(\cdot)d\mu(t)\right)(\widehat{x}).$$

12.2. The transport and Monge–Kantorovich problems. Suppose some homogeneous commodity has to be delivered from the sites where it is produced to the sites of its consumption. Let the vector $c = (c_1, \dots, c_m)$ represent the amounts of production and consumption at m sites (the amounts of consumption are positive, while the amounts of production are nonpositive, the produced quantity being given by the absolute value of the corresponding component). Assume that the total amount of consumption equals the total amount of production, i.e., $\sum_{k=1}^m c_k = 0$. The transportation plan is described by a matrix (x_{ij}), $1 \le i, j \le m$, $x_{ij} \ge 0$, whose generic element x_{ij} determines the planned amount of transportation from site i to site j. The balance relations are given by the equalities $\sum_{i=1}^m x_{ik} - \sum_{j=1}^m x_{kj} = c_k$, $1 \le k \le m$, which means that, for each k, $\sum_{i=1}^m x_{ik}$ units of the commodity are delivered to and $\sum_{j=1}^m x_{kj}$ units are taken from the kth site, and these quantities must amount to c_k units.

Denote by a_{ij} the transportation cost of a unit of goods from the ith to the jth site. Then minimization of the total transportation cost leads us to the following extremal problem:

$$\sum_{i,j=1}^m a_{ij}x_{ij} \to \min, \quad \sum_{i=1}^m x_{ik} - \sum_{j=1}^m x_{kj} = c_k, \qquad 1 \le k \le m,\ x_{ij} \ge 0. \tag{P}$$

The problems of this kind were first studied by the Soviet economist A. N. Tolstoy in the 30s and American mathematician F. Hitchcock in the early 40s.

The problem (P) is a linear programming problem. It is obviously consistent; i.e., it has a solution. Its value is identical to the value of the dual problem

$$\sum_{k=1}^m y_k c_k \to \max, \quad y_i - y_j \le a_{ij} \qquad 1 \le i, j \le m.$$

This problem also has a solution which determines what is to be transported from one site to another (the y_i are Lagrange multipliers, and transportation from i to j is done if $y_i - y_j = a_{ij}$). Analyzing the initial and dual problems simultaneously allows one to find the optimal transportation plan.

Nowadays effective algorithms are available for the practical solution of the transportation problem.

Actually the first linear programming problem was posed and studied by Gaspare Monge (1784). Monge's problem, a continuous analog of the transport problem, consists in finding the most rational way to carry earth from one place (making a moat) to another (to build a rampart). Mathematically this problem is posed as follows: given two planar or spacial figures (bodies) of equal area (volume) partitioned into infinitesimal elements, establish a correspondence between the elements of the two figures (bodies) to minimize the sum of products of the distances between the corresponding elements by their area (volume) (i.e., to minimize the work needed to carry one figure (body) into the other piece by piece).

In modern terms the problem may be formulated as follows. Let $X = (X, d)$ be a complete (separable) metric space with Borel σ-algebra $\mathcal{B}(X)$, let $a\colon X \times X \to \mathbb{R}$ be the cost function and c a Borel measure on X. Then the problem is:

$$\int_{X\times X} a(x,y)\mu(d(x,y)) \to \min, \quad \mu(B \times X) - \mu(X \times B) = c(B), \quad \forall B \in \mathcal{B}(X), \tag{P_1}$$

with minimization over all countably additive measures μ on $X \times X$ of finite total variation.

Monge considered the distance between the corresponding points as the cost and minimized it under given projections, $\mu(B \times X) = \mu_1(B)$ and $\mu(X \times B) = \mu_2(B)$ (of course, his formulation did not involve any measures and metric spaces). The problem (P_1) was posed by Kantorovich (1942) and then was treated by Kantorovich and Rubinstein (1957). Afterwards this problem became known as the *Monge–Kantorovich problem*.

The theory underlying the investigation of problem (P_1) consists of two ingredients: the duality theory describing the cost functions a

such that the value $S(a,c)$ of problem (P_1) is given by

$$S(a,c) = \sup\left\{ \int_X u(x)c(dx) \,\middle|\, u(\cdot) \in C^b(X),\ u(x) - u(y) \le a(x,y),\ \forall x, y \in X \right\}$$

(with $C^b(X)$ being the space of bounded continuous functions on X), and the description of the Kantorovich–Rubinstein metric $K_{\mathrm{KR}}(\mu_1, \mu_2) = S(c, \mu_1 - \mu_2)$.

For a further development of this subject see Levin (1978) and Levin and Milyutin (1979).

12.3. Convexity and differentiability.

Obviously, a convex function may not be differentiable at some points, but it is differentiable on massive sets.

THEOREM 1 (RADEMACHER). *A finite convex function on a finite-dimensional space is differentiable almost everywhere.*

THEOREM 2 (MASUR). *Let X be a separable Banach space and A an open convex subset of X. If a function $f\colon A \to \mathbb{R}$ is convex and continuous, then it is Gâteaux differentiable on a dense G_δ-subset of A.*

COROLLARY. *The collection of smoothness points of the boundary of a convex open set is a dense G_δ-subset of the boundary.*

Problems of this kind attracted the attention of many mathematicians, and a number of beautiful results were obtained. Here we mention one of them. A Banach space is said to be an *Asplund space* if any convex continuous function on a convex open set in this space is Frechét differentiable on a dense G_δ-subset of this set.

THEOREM 3. *A Banach space X is an Asplund space if and only if one of the following conditions is fulfilled*: (*a*) *any equivalent norm in X is Frechét differentiable at least at one point* (*then it is differentiable on a dense G_δ-subset*); (*b*) *any separable subspace of X has a separable dual.*

There is an extensive literature on problems of this type. We point out the monographs by Phelps (1978) and Giles (1982).

12.4. Convexity in the theory of multi-valued mappings.

We will state here only two results of principal importance in the theory of multi-valued mappings in which convexity plays an essential role.

THEOREM (Michael's selection theorem). *Let T be a paracompact topological space, X a Banach space, and $F\colon T \to X$ a lower semicontinuous multi-valued mapping such that, for any $t \in T$, $F(t)$ is a nonempty closed convex subset of X. Then there exists a continuous function $f\colon T \to X$ such that $f(t) \in F(t)$ for any $t \in T$.*

THEOREM (Tychonoff–Fan–Kakutani fixed-point theorem). *Let X be a locally convex space, A a nonempty convex compact set in X, and $F\colon A \to A$ an upper semicontinuous multi-valued mapping. Then there exists a point $x_0 \in A$ such that $x_0 \in F(x_0)$.*

For a detailed exposition the reader is referred to Holmes (1975).

12.5. Convexity and ordering.

Convexity is closely connected with order relation on the set of real numbers. This is what enables us to define the notions of convex set and convex function. Attempting to extend convex analysis to operators, it is natural to begin with spaces in which an order relation is defined. The most suitable spaces in this respect are the so-called K-spaces, or the *Kantorovich spaces*. A K-space is an ordered vector space (see Section 2.1) in which there exist the upper and lower bounds of any two elements and any nonempty bounded set also has its upper and lower bounds. The K-spaces admit a result completely analogous to the Hahn–Banach theorem.

THEOREM (Hahn–Banach–Kantorovich). *Let X be a vector space, Y a K-space, X_0 a subspace of X, $P\colon X \to Y$ a sublinear operator, and $\Lambda_0\colon X_0 \to Y$ a linear operator such that $\Lambda_0(x) \le P(x)$ for all $x \in X_0$. Then there is a linear operator $\Lambda\colon X \to Y$ such that $\Lambda(x) = \Lambda_0(x)$ if $x \in X_0$ and $\Lambda(x) \le P(x)$ for all $x \in X$.*

Actually the only nontrivial sublinear operator in the theory of convex functions is the operator of taking supremum. The reason is that any closed sublinear function is a supremum of linear ones. A similar property holds for sublinear operators with values in K-spaces. Let Y be a K-space and $\mathcal{U}$ an arbitrary nonempty set. Denote by $l_\infty(\mathcal{U}, Y)$ the class of all bounded mappings from $\mathcal{U}$ into Y. The operator $\mathcal{E}_\mathcal{U}\colon l_\infty(\mathcal{U}, Y) \to Y$ defined by the formula $\mathcal{E}_\mathcal{U} f = \sup\{f(A) \mid A \in \mathcal{U}\}$ is the *Kutateladze canonical operator*. Any sublinear operator defined everywhere on $\mathcal{U}$ is obtained as a superposition of the canonical operator with a linear one. Thus to develop the subdifferential calculus one has to study only the properties of the canonical operator. We refer the reader to Kutateladze (1979) for details and further references.

12.6. Convex integer-valued polyhedra.

In the mid-70s surprising connections between the theory of integer-valued polyhedra and solvability of systems of polynomial equations were discovered.

A function $f(z) = \sum_{k\in\mathbb{Z}^n} a_k z^k$, where $z \in (\mathbb{C} \setminus \{0\})^n$ and only finitely many coefficients $a_k \in \mathbb{Z}^n$ are nonzero, is called a *Laurent polynomial.* The set $\mathbb{Z}^n_f = \{k \in \mathbb{Z}^n \mid a_k \neq 0\}$ is called the support of f. The convex hull of $\mathbb{Z}^n_f$ is referred to as the *Newton polyhedron*, written $\Gamma(f)$. Let f be a Laurent polynomial. How many solutions (in a general position) has the equation $f'(z) = 0$? Let $\{f_i\}_{i=1}^n$ be n Laurent polynomials. How many solutions has the system of equations $f_i(z) = 0$, $i = 1, \dots, n$? The answers are given in terms of convex analysis.

Let $\mathcal{A} = \{A_i\}_{i=1}^n$ be a system of n convex bodies in $\mathbb{R}^n$ and let $V(A)$ denote the volume of a set A. The mixed volume of the system $\mathcal{A}$ is the number

$$V(\mathcal{A}) = (-1)^{n-1}\sum_{i=1}^{n} V(A_i) + (-1)^{n-2} \times \sum_{i<j} V(A_i + A_j) + \cdots + V(A_1 + \cdots + A_n).$$

THEOREM 1 (Koushnirenko (1976)). *The number of critical points of a Laurent polynomial (i.e.,* card$\{z \mid f'(z) = 0\}$*) in a general position equals $V(\Gamma_f)$ (the volume of the Newton polynomial).*

THEOREM 2 (Khovansky, Bernstein, and Kushnirenko (1976)). *The number of solutions to the system of equations $f_i(z) = 0$, $1 \leq i \leq n$, in a general position equals the mixed volume of the system of sets $\{\Gamma_{f_i}\}_{i=1}^n$.*

This implies, in particular, that this mixed volume is integer-valued!

Note that the well-known Alexandroff–Fenchel inequalities for mixed volumes are remote generalizations of the isoperimetric inequality. Quite recently close connections between the Alexandroff–Fenchel inequalities and algebraic geometry were discovered (see Burago and Zalgaller (1988)).

13. Convex analysis and the theory of extremum

13.1. The Lagrange principle for smooth-convex problems.

The role of convex analysis in the theory of extremum is not limited to the study of convex problems. Many extremal problems combine the smoothness and convexity structures. The aim of this section is to justify the following assertion: *the problems involving the smoothness and convexity structures admit a necessary condition for extremum corresponding to the Lagrange principle of eliminating constraints according to which a necessary condition for extremum in a problem with constraints coincides with necessary conditions for the Lagrange function "as if the variables were independent"* (quoted from Lagrange; we will somewhat strengthen this statement). Now we proceed to exact formulations.

Let X and Y be normed spaces, U_0 an open subset of X, and $\mathcal{U}$ a set; and let a functional $f_0\colon U_0 \to \mathbb{R}$ and a mapping $F\colon U_0 \times \mathcal{U} \to Y$ be given. Consider the problem

$$f_0(x) \to \min, \qquad F(x,u) = 0, \quad u \in \mathcal{U}. \tag{P}$$

If the functional f_0 and the mapping F possess certain smoothness in x and the mapping certain convexity properties in u, we say that (P) is a *smooth-convex problem.*

A pair $(\widehat{x}, \widehat{u}) \in U_0 \times \mathcal{U}$ is said to afford a *strong local minimum* in problem (P) if there is a $\delta > 0$ such that for any pair $(x,u) \in U_0 \times \mathcal{U}$ satisfying the conditions $F(x,u) = 0$ and $\|x - \widehat{x}\|_X < \delta$ the inequality $f_0(x) \ge f_0(\widehat{x})$ holds.

The function

$$\mathcal{L}\left((x,u), \lambda_0, \lambda\right) = \lambda_0 f_0(x) + \langle \lambda, F(x,u)\rangle$$

is referred to as the *Lagrange function of problem* (P), and the number λ_0 and the functional $\lambda \in Y^*$ as the *Lagrange multipliers.*

Recall the definitions of (Fréchet) differentiability and strict differentiability of a mapping. Let X and Y be normed spaces and U

a neighborhood of a point $\widehat{x} \in X$. A mapping $\Phi\colon U \to Y$ is said to be *differentiable at the point* $\widehat{x}$ if there is a linear continuous operator $\Lambda\colon X \to Y$ such that for all $h \in X$ for which $\widehat{x} + h \in U$ the following equality holds:

$$\Phi(\widehat{x} + h) = \Phi(\widehat{x}) + \Lambda h + r(h),$$

where $r(h) = o(h)$, i.e., $\lim_{h\to 0}(\|r(h)\|_Y/\|h\|_X) = 0$. The operator Λ is the *Fréchet derivative of the mapping* Φ *at the point* $\widehat{x}$, denoted by $\Phi'(\widehat{x})$.

A mapping Φ is said to be *strictly differentiable at a point* $\widehat{x}$ if it is Fréchet differentiable at this point and for any $\varepsilon > 0$ there is a $\delta > 0$ such that

$$\|\Phi(x_1) - \Phi(x_2) - \Phi'(\widehat{x})(x_1 - x_2)\|_Y \leq \varepsilon\|x_1 - x_2\|_X$$

whenever $\|x_i - \widehat{x}\|_X < \delta$, $i = 1, 2$.

THEOREM (Lagrange principle for smooth-convex problems). *Let X and Y be Banach spaces, U a neighborhood of a point $\widehat{x} \in X$, $\mathcal{U}$ a set, and $\widehat{u} \in \mathcal{U}$. Let f_0 be a function on U differentiable at $\widehat{x}$ and $F\colon U \times \mathcal{U} \to Y$ a mapping such that $x \to F(x, \widehat{u})$ is strictly differentiable at $\widehat{x}$ and the sets $F(x, \mathcal{U})$ are convex in Y for all $x \in U$. Suppose the subspace $\operatorname{Im} F_x(\widehat{x}, \widehat{u})$ is closed and has a finite codimension.*

If $(\widehat{x}, \widehat{u})$ is a strong local minimum point of problem (P), then at this point the Lagrange principle holds; i.e., there are Lagrange multipliers $\lambda_0 \geq 0$ and $\lambda \in Y^*$ not all equal to zero such that the following conditions are fulfilled:

(*a*) *the stationarity conditions at the point $\widehat{x}$ in the problem without constraints $\mathcal{L}((x, \widehat{u}), \lambda_0, \lambda) \to \min$, $x \in X$, i.e.,*

$$\mathcal{L}_x((\widehat{x}, \widehat{u}), \lambda_0, \lambda) = 0 \quad \Longleftrightarrow \quad \lambda_0 f_0'(\widehat{x}) + (F_x(\widehat{x}, \widehat{u}))^* \lambda = 0;$$

(*b*) *the minimum principle*

$$\min_{u \in \mathcal{U}} \mathcal{L}((\widehat{x}, u), \lambda_0, \lambda) = \mathcal{L}((\widehat{x}, \widehat{u}), \lambda_0, \lambda).$$

The proof relies on three results which we will state without proof (which can be found in Alekseev, Tikhomirov, and Fomin (1979), pp. 173, 130).

Let M be a subset of a normed space X and $x_0 \in M$. An element $h \in X$ is a *tangent vector to M at x_0* if there exist an $\varepsilon > 0$ and a mapping $r\colon (-\varepsilon, \varepsilon) \to X$ such that $x_0 + th + r(t) \in M$ for all $t \in (-\varepsilon, \varepsilon)$ and $r(t) = o(t)$ as $t \to 0$.

Denote the set of all tangent vectors to M at the point x_0 by $T_{x_0}M$.

THEOREM (Lyusternik). *Let X, Y be Banach spaces and U a neighborhood of a point $x_0 \in X$. Let $\Phi\colon U \to Y$ be a mapping strictly differentiable at x_0, let $\Phi'(x_0)\colon X \to Y$ be surjective, let $\Phi(x_0) = 0$, and let $M = \{x \in X \mid \Phi(x) = 0\}$. Then*

$$T_{x_0}M = \operatorname{Ker} \Phi'(x_0).$$

LEMMA (on the annihilator of the kernel of a regular operator). *Let X, Y be Banach spaces, let $\Lambda\colon X \to Y$ be a linear continuous surjective operator and Λ^* its conjugate. Then*

$$(\operatorname{Ker} \Lambda)^{\perp} = \operatorname{Im} \Lambda^*.$$

Recall that a hyperplane *separates one set from another* if these sets lie in different half-spaces relative to this hyperplane. The following result is a consequence of separation theorems in Section 1.

THEOREM (finite-dimensional separation theorem). (a) *A point which is not interior to a convex set in a finite-dimensional space can be separated from it by a hyperplane.*

(b) *Two disjoint convex sets in a finite-dimensional space can be separated from each other by a hyperplane.*

PROOF OF THE LAGRANGE PRINCIPLE. The notion of smoothness is related to analysis, whereas that of convexity is related to geometry. The proof of the Lagrange principle will consist of three parts, one of which is analytic, and the other two geometric, based on the finite-dimensional separation theorem.

Denote $\Lambda = F_x(\widehat{x}, \widehat{u})$, $C = F(\widehat{x}, \mathcal{U})$, $Y_1 = \operatorname{Im} \Lambda$, $Z = Y/Y_1$ (by condition, $\dim Z < \infty$); and let $\pi\colon Y \to Z$ be the canonical projection. We will distinguish between two cases: degenerate, where either $\operatorname{int} \pi(C) = \emptyset$ or $\operatorname{int} \pi(C) \neq \emptyset$ but $0 \notin \operatorname{int} \pi(C)$, and nondegenerate, where $\operatorname{int} \pi(C) \neq \emptyset$ and $0 \in \operatorname{int} \pi(C)$.

Degenerate case. Here we use geometry. By part a) of the finite-dimensional separation theorem there is a hyperplane H in Z separating 0_Z from $\pi(C)$. The inverse image $\pi^{-1}H$ of this hyperplane is a hyperplane $\mathcal{H} \in Y$ passing through the origin and such that C lies in one of the half-spaces determined by $\mathcal{H}$; i.e., there is an element $\lambda \in Y^*$ such that $\mathcal{H} = \operatorname{Ker} \lambda$ and $\langle \lambda, x \rangle \geq 0 \ \forall x \in C$. Then, since $\operatorname{Im}\Lambda \subset \operatorname{Ker}\lambda$, we have $\langle \mathcal{L}_x((\widehat{x},\widehat{u}),0,\lambda),x\rangle = \langle \lambda, \Lambda x\rangle = 0$ for any $x \in X$, i.e., $\Lambda^*\lambda = 0$, which proves (a) with $\lambda_0 = 0$.

Next, if $u \in \mathcal{U}$, we obtain $\mathcal{L}((\widehat{x},u),0,\lambda) = \langle \lambda, F(\widehat{x},u)\rangle \geq 0$ (since $\langle \lambda, x\rangle \geq 0\ \forall x \in C$), which proves (b) for $\lambda_0 = 0$.

Nondegenerate case. Let $0 \in \operatorname{int}\pi(C)$. Then, obviously, $\operatorname{span}\pi(C) = Z$. Let $e_1,\dots,e_m$ be a basis in Z; then, obviously, $\operatorname{cone}\{\pm e_i, 1 \leq i \leq m\} = \operatorname{span}\pi(F(\widehat{x},\mathcal{U}))$. It is clear that there is $\beta > 0$ such that $\pm\beta e_i \in \pi(C)$, $i = 1,\dots,m$. Denote the set of all of these vectors by $\{w_j\}_{j=1}^{2m}$. Clearly, $\sum_{j=1}^{2m} w_j = 0$. Let $v_j \in \mathcal{U}$ be vectors such that $w_j = \pi(F(\widehat{x},v_j))$, $j = 1,\dots,2m$. Then $\pi(\sum_{j=1}^{2m} F(\widehat{x},v_j)) = \sum_{j=1}^{2m} \pi(F(\widehat{x},v_j)) = 0$. Therefore there exists a vector $\xi \in X$ such that

$$\Lambda\xi + \sum_{j=1}^{2m} F(\widehat{x},v_j) = 0. \tag{i}$$

Let $y \in Y_1 \cap C$. Then there are $x_y \in X$ and $v_y \in \mathcal{U}$ such that

$$\Lambda x_y + F(\widehat{x},v_y) = 0. \tag{ii}$$

For an arbitrary $\varepsilon > 0$ consider the mapping $\Phi_\varepsilon \colon U \times \mathbb{R} \times \mathbb{R}^{2m} \to Y$,

$$\begin{aligned}\Phi_\varepsilon(x,\alpha,z) = &\Big(1 - \alpha(1+\varepsilon 2m) - \sum_{j=1}^{2m} z_j\Big)F(\widehat{x}+x,\widehat{u}) + \alpha F(\widehat{x}+x,v_y)\\ &+ \alpha\varepsilon\sum_{j=1}^{2m} F(\widehat{x}+x,v_j) + \sum_{j=1}^{2m} z_j F(\widehat{x}+x,v_j).\end{aligned} \tag{iii}$$

It is easily seen that this mapping is strictly differentiable at the point $(0,0,0)$ with the derivative

$$\Phi'_\varepsilon(0,0,0)[x,\alpha,z] = \Lambda x + \alpha F(\widehat{x},v_y) + \alpha\varepsilon\sum_{j=1}^{2m} F(\widehat{x},v_j) + \sum_{j=1}^{2m} z_j F(\widehat{x},v_j),$$

and the operator $\Phi'_\varepsilon(0,0,0)\colon X \times \mathbb{R} \times \mathbb{R}^{2m} \to Y$ is surjective. Moreover, the last formula, (i), and (ii) imply that $\Phi'_\varepsilon(0,0,0)[x_y+\varepsilon\xi,1,0] = 0$. Therefore by Lyusternik's theorem $(x_y+\varepsilon\xi,1,0)$ is a tangent vector to the set $\{(x,\alpha,z) \mid \Phi_\varepsilon(x,\alpha,z) = 0\}$ at the point $(0,0,0)$; hence there is a $\delta > 0$ and functions $r_{y\varepsilon}\colon (-\delta,\delta) \to X$, $\rho_{y\varepsilon}\colon (-\delta,\delta) \to \mathbb{R}$, and $z_{y\varepsilon}\colon (-\delta,\delta) \to \mathbb{R}^{2m}$ such that each of them is $o(t)$ as $t \to 0$ and for any $t \in (-\delta,\delta)$ they satisfy the equality

$$\Phi_\varepsilon(t(x_y+\varepsilon\xi) + r_{y\varepsilon}(t), t + \rho_{y\varepsilon}(t), z_{y\varepsilon}(t)) = 0.$$

As follows from (iii), for any sufficiently small $t > 0$ the left-hand side of the last equality is a convex combination of elements of the set $F(t(x_y + \varepsilon\xi) + r_{y\varepsilon}(t), \mathcal{U})$. Since this set is convex, there exists its element $u_{y\varepsilon t}$ such that $F(\widehat{x} + t(x_y + \varepsilon\xi) + r_{y\varepsilon}(t), u_{y\varepsilon t}) = 0$. Thus $\widehat{x} + t(x_y + \varepsilon\xi) + r_{y\varepsilon}(t)$ is an admissible point in the problem (P); hence for all such t

$$f_0(\widehat{x} + t(x_y + \varepsilon\xi) + r_{y\varepsilon}(t)) = f_0(\widehat{x}) + t\langle f_0'(\widehat{x}), x_y + \varepsilon\xi\rangle + o(t) \geq f_0(\widehat{x}).$$

Hence, dividing by t and taking the limit, we obtain $\langle f_0'(\widehat{x}), x_y\rangle \geq -\varepsilon\langle f_0'(\widehat{x}), \xi\rangle$, and consequently, since $\varepsilon > 0$ is arbitrary,

$$\langle f_0'(\widehat{x}), x_y\rangle \geq 0. \qquad (iv)$$

Take $y = 0 \in Y_1 \cap C$. Then relation (ii) holds, in particular, for $v_0 = \widehat{u}$ and any $x_0 \in \operatorname{Ker}\Lambda$. It follows from (iv) that the functional $f_0'(\widehat{x})$ is nonnegative on the subspace $\operatorname{Ker}\Lambda$; hence it is equal to zero, i.e., $f_0'(\widehat{x}) \in (\operatorname{Ker}\Lambda)^\perp$. By condition, Y_1 is a closed subspace of Y; hence it is a Banach space. By the lemma on the annihilator of the kernel of a regular operator (applied to X and Y_1) there is $y_1^* \in Y_1^*$ such that

$$f_0'(\widehat{x}) + \Lambda^* y_1^* = 0. \qquad (v)$$

Let again $y \in Y_1 \cap C$ and let x_y and v_y satisfy (ii). Then, as we proved,

$$\langle y_1^*, F(\widehat{x}, v_y)\rangle = -\langle y_1^*, \Lambda x_y\rangle = \langle \Lambda^* y_1^*, x_y\rangle = \langle f_0'(\widehat{x}), x_y\rangle \geq 0. \qquad (vi)$$

We will show that there exists an extension λ of the functional y_1^* to the entire Y such that $\langle \lambda, F(\widehat{x}, u)\rangle \geq 0$ for all $u \in \mathcal{U}$ (which will mean that (b) holds with $\lambda_0 = 1$). To this end we again use geometry. The sets $\Pi = \{y_1 \in Y_1 \mid \langle y_1^*, y_1\rangle < 0\}$ and C are obviously disjoint. Consider the canonical projection $\pi_1 \colon Y \to Y/Y_2$, where $Y_2 = Y_1 \cap \operatorname{Ker} y_1^*$. Then the sets $\pi_1\Pi$ and $\pi_1 C$ are also disjoint, since otherwise there would be elements $\xi_1 \in \Pi$ and $\xi_2 \in C$ such that $\xi_1 - \xi_2 \in Y_2$ (so that $\xi_2 \in Y_1 \cap C$); hence $\langle y_1^*, \xi_1\rangle = \langle y_1^*, \xi_2\rangle$. But the left-hand side of this equality is negative by the definition of Π, whereas the right-hand side is nonnegative by (vi). The space Y/Y_2 is finite-dimensional; hence by part (b) of the separation theorem the sets $\pi_1\Pi$ and $\pi_1 C$ can be separated by a hyperplane H_1. The inverse image $\pi^{-1}H_1$ of this hyperplane is a hyperplane $\mathcal{H}_1 \in Y$ such that its trace on Y_1 separates Π and $C \cap Y_1$. A linear functional $\lambda \in Y^*$ with kernel $\mathcal{H}_1$ is defined up to a constant factor. Define this factor so that this functional would coincide with y_1^* on Y_1. Then it is nonnegative on C, and we have

$0 = \langle f_0'(\widehat{x}), x\rangle + \langle y_1^*, \Lambda x\rangle = \langle f_0'(\widehat{x}), x\rangle + \langle y^*, \Lambda x\rangle$, which implies the stationarity condition with $\lambda_0 = 1$, while the nonnegativity condition $\langle \lambda, F(\widehat{x}, u)\rangle \geq 0\ \forall u \in \mathcal{U}$ implies the maximum principle. □

13.2. Applications.

In this subsection we obtain necessary conditions for extremum in several extremal problems as a consequence of the Lagrange principle proved in the previous part. To this end we need some preliminaries to be stated as lemmas (the proofs can be found in Alekseev, Tikhomirov, and Fomin (1979)).

Recall that for an interval Δ of the real line and $m \geq 1$ we write $C(\Delta, \mathbb{R}^m)$ and $C^1(\Delta, \mathbb{R}^m)$ for Banach spaces of continuous and continuously differentiable vector functions on Δ respectively.

LEMMA 1 (on nontriviality of the annihilator). *Let X be a normed space and L its closed proper subspace. Then $L^\perp \neq \{0_X\}$.*

LEMMA 2 (on differentiabilily of integral functionals). *Let $\Delta = [t_0, t_1]$ be an interval of the real line, V an open subset of $\mathbb{R} \times \mathbb{R}^n \times \mathbb{R}^r$ (in variables (t, x, u), $t \in \mathbb{R}$, $x \in \mathbb{R}^n$, and $u \in \mathbb{R}^r$), let $\widehat{x}(\cdot) \in C^1(\Delta, \mathbb{R}^n)$ and $\widehat{u}(\cdot) \in C(\Delta, \mathbb{R}^r)$, and suppose that the graph of $\{(t, \widehat{x}(t), \widehat{u}(t)) \mid t \in \Delta\}$ belongs to V. Then if a function $f\colon V \to \mathbb{R}$ is continuous on V together with its partial derivatives with respect to x and u, the functional $J(z) = \int_{t_0}^{t_1} f(t, x(t), u(t))\, dt$ defined on the set $W = \{z = (x(\cdot), u(\cdot)) \in C^1(\Delta, \mathbb{R}^n) \times C(\Delta, \mathbb{R}^r) \mid (t, x(t), u(t)) \in V,\ t \in \Delta\}$ is continuously differentiable at the point $\widehat{z} = (\widehat{x}(\cdot), \widehat{u}(\cdot))$ and*

$$\langle J'(\widehat{z}), z\rangle = \int_{t_0}^{t_1} (\widehat{f}_x(t) \cdot x(t) + \widehat{f}_u(t) \cdot u(t))\, dt,$$

where $\widehat{f}_x(t) = f_x(t, \widehat{x}(t), \widehat{u}(t))$ and $\widehat{f}_u(t) = f_u(t, \widehat{x}(t), \widehat{u}(t))$.

LEMMA 3 (on differentiability of a differential constraint). *Let Δ and V be as in Lemma 2 and let the graph of $\{(t, \widehat{x}(t), \widehat{u}(t)) \mid t \in \Delta\}$ belong to V. If the mapping $\varphi\colon V \to \mathbb{R}^n$ and its partial derivatives with respect to x and u are continuous on V, then the mapping $\mathcal{F}\colon W \to \mathbb{R}^n$, $\mathcal{F}(z)(t) = \dot{x}(t) - \varphi(t, x(t), u(t))$, with W as in Lemma 2, is continuously differentiable at $\widehat{z} = (\widehat{x}(\cdot), \widehat{u}(\cdot))$ and*

$$\mathcal{F}'(\widehat{z})[z](t) = \dot{x}(t) - \widehat{\varphi}_x(t)x(t) - \widehat{\varphi}_u(t)u(t),$$

where $\widehat{\varphi}_x(t) = \varphi_x(t, \widehat{x}(t), \widehat{u}(t))$ and $\widehat{\varphi}_u(t) = \varphi_u(t, \widehat{x}(t), \widehat{u}(t))$.

LEMMA 4 (on strict differentiability). *Let X and Y be normed spaces, U a neighborhood of a point $\widehat{x} \in X$, and suppose that $F\colon U \to Y$ is a mapping continuously differentiable on U. Then F is strictly differentiable at $\widehat{x}$.*

LEMMA 5 (on closedness of the image). *Let X, Y, and Z be Banach spaces, let $A\colon X \to Y$, $B\colon X \to Z$, and let $T = A \times B\colon X \to Y \times Z$, $Tx = (Ax, Bx)$ be linear operators. If the subspace $\operatorname{Im} A$ is closed in Y and the subspace $B(\operatorname{Ker} A)$ is closed in Z, then the subspace $\operatorname{Im} T$ is closed in $Y \times Z$.*

LEMMA 6 (on existence of a solution to a linear system). *Let Δ be an interval, M^{mn} the space of $m \times n$ matrices, $A(\cdot)\colon \Delta \to M^{mn}$ a piecewise continuous matrix function, and $y(\cdot) \in C(\Delta, \mathbb{R}^n)$. Then for any $\tau \in \Delta$ and any $\xi \in \mathbb{R}^n$ there exists a solution to the Cauchy problem $\dot{x} = A(t)x + y(t)$, $x(\tau) = \xi$, defined on the entire Δ.*

The next result is an appropriate version of the theorem by A. A. Lyapunov proved in Section 11.

LEMMA 7 (Lyapunov's theorem). *Let $a(\cdot) = (a_1(\cdot), \dots, a_n(\cdot))$ be an integrable vector function on an interval Δ and let Σ be the class of all measurable subsets of Δ. Then $\{x \in \mathbb{R}^n \mid \exists\, B \in \Sigma\colon x = \int_B a(t)dt\}$ is a convex compact set in $\mathbb{R}^n$.*

Now we proceed to the treatment of specific extremal problems. We will refer to the Lagrange principle for smooth-convex problems as the smooth-convex principle. The Lagrange principle have been proved for maximization problems, but in a minimization problem one can change the sign of the functional and consider the corresponding maximization problem (with the same constraints), which will clearly have the same value. For this reason some problems are (traditionally) considered for extremum (i.e., simultaneously for maximum and minimum).

1. *Problems without constraints.*

COROLLARY 1 (Fermat's theorem). *Let X be a Banach space, let U be a neighborhood of a point $\widehat{x} \in X$, and let $f\colon U \to \mathbb{R}$ be a differentiable function at $\widehat{x}$. If $\widehat{x}$ affords a local extremum in the problem without constraints*

$$f(x) \to \text{extr}, \tag{P_1}$$

then the stationarity condition holds:

$$f'(\widehat{x}) = 0. \tag{1}$$

The proof follows from the smooth-convex principle in an obvious way. But a direct proof of this fact is also elementary and does not require that X be a Banach space.

REMARK. A prototype of relation (1) for functions of a single variable was first formulated by P. Fermat in 1638.

2. *Smooth problems with equality-type constraints.*

COROLLARY 2 (the Lagrange multipliers rule). *Let X and Y be Banach spaces, let U be a neighborhood of a point $\widehat{x} \in X$, and let $f\colon U \to \mathbb{R}$ be a differentiable function and $F\colon U \to Y$ a strictly differentiable mapping at $\widehat{x}$. Further, let the subspace* $\operatorname{Im} F'(\widehat{x})$ *be closed in Y. Then if $\widehat{x}$ affords a local extremum in the problem with equality-type constraints,*

$$f_0(x) \to \text{extr}, \qquad F(x) = 0, \tag{P_2}$$

this problem obeys the Lagrange principle: *there are Lagrange multipliers $\lambda_0 \in \mathbb{R}$ and $\lambda \in Y^*$ not equal to zero simultaneously such that the Lagrange function $\mathcal{L}(x, \lambda_0, \lambda) = \lambda_0 f_0(x) + \langle \lambda, F(x) \rangle$ of the problem (P_2) satisfies the stationarity condition*:

$$\mathcal{L}_x(\widehat{x}, \lambda_0, \lambda) = 0 \iff \lambda_0 \langle f_0'(\widehat{x}), x \rangle + \langle \lambda, F'(\widehat{x}) x \rangle = 0, \ \forall x \in X. \tag{2}$$

PROOF. If $\operatorname{Im} F'(\widehat{x})$ is not the entire Y, then by Lemma 1 there exists a nonzero functional $\lambda \in Y^*$ such that $\langle \lambda, F'(\widehat{x}) x \rangle = 0$ for all $x \in X$; i.e., (2) holds with $\lambda_0 = 0$. If $\operatorname{Im} F'(\widehat{x}) = Y$, then we apply the smooth-convex principle with the set $\mathcal{U}$ consisting of a single point to obtain (2). □

REMARK. In the setting where $X = \mathbb{R}^n$ and $Y = \mathbb{R}^m$ the Lagrange multipliers rule was first formulated in 1797 by J. Lagrange. Its proof was given only in the second half of the 19th century. Corollary 2 in the regular case where $\operatorname{Im} F'(\widehat{x}) = Y$ was proved by L. A. Lyusternik (1934).

3. *The problems of classical calculus of variations.*

COROLLARY 3 (the Lagrange principle for the Lagrange problem with fixed time). *Let V and W be open sets in the spaces $\mathbb{R}\times\mathbb{R}^n\times\mathbb{R}^r$ and $\mathbb{R}^n\times\mathbb{R}^n$ respectively; let functions $f_i\colon V\to\mathbb{R}$, $i=0,1,\dots,m$, and a mapping $\varphi\colon V\to\mathbb{R}^n$ (of variables (t,x,u), $t\in\mathbb{R}, x\in\mathbb{R}^n, u\in\mathbb{R}^r$), along with their partial derivatives in x and u, be continuous on V; and let functions $\psi_i\colon W\to\mathbb{R}$, $i=0,1,\dots,m$ (of variables (ξ_0,ξ_1), $\xi_j\in\mathbb{R}^n$, $j=0,1$) be continuously differentiable on W. Further, let $\Delta=[t_0,t_1]$ and $\widehat{x}(\cdot)\in C^1(\Delta,\mathbb{R}^n)$, $\widehat{u}(\cdot)\in C(\Delta,\mathbb{R}^r)$ be such that $\{(t,\widehat{x}(t),\widehat{u}(t))\mid t\in\Delta\}\subset V$, $(\widehat{x}(t_0),\widehat{x}(t_1))\in W$. If $\widehat{z}=(\widehat{x}(\cdot),\widehat{u}(\cdot))\in C^1(\Delta,\mathbb{R}^n)\times C(\Delta,\mathbb{R}^r)$ affords a local minimum in the Lagrange problem ($z=(x(\cdot),u(\cdot))$)*

$$\mathcal{B}_0(z)=\int_{t_0}^{t_1} f_0(t,x(t),u(t))\,dt+\psi_0(t_0,x(t_0),t_1,x(t_1))\to\min,$$

$$\mathcal{B}_i(z)=\int_{t_0}^{t_1} f_i(t,x(t),u(t))\,dt+\psi_i(t_0,x(t_0),t_1,x(t_1))=0, \qquad (P_3)$$

$$i=1,\dots,m,$$

$$\dot{x}=\varphi(t,x(t),u(t)),$$

then to this problem the Lagrange principle applies: there exist Lagrange multipliers $p(\cdot)\in C^1(\Delta,(\mathbb{R}^n)')$ and $\mu_0,\mu_1,\dots,\mu_m$ not equal to zero simultaneously such that the Lagrange function

$$\mathcal{L}(z,p(\cdot),\mu_0,\dots,\mu_m)=\int_{t_0}^{t_1} L(t,x(t),\dot{x}(t),u(t))\,dt+l(x(t_0),x(t_1)),$$

where $L=g+p\cdot(\dot{x}-\varphi)$, $g=\sum_{i=0}^m\mu_i f_i$, $l=\sum_{i=0}^m\mu_i\psi_i$, satisfies

— *the stationarity conditions in x (Euler's equation) and u:*

$$-\frac{d}{dt}\widehat{L}_{\dot{x}}(t)+\widehat{L}_x(t)=0 \iff -\dot{p}(t)=p(t)\cdot\widehat{\varphi}_x(t)-\widehat{g}_x(t); \qquad (3a)$$

$$\widehat{L}_u(t)=0 \iff p(t)\cdot\widehat{\varphi}_u(t)=\widehat{g}_u(t), \qquad (3b)$$

where $\widehat{L}_{\dot{x}}(t)=L_{\dot{x}}(t,\widehat{x}(t),\dot{\widehat{x}}(t),\widehat{u}(t))$ with similar definitions for other functions with a "hat",

— *and the transversality conditions in x:*

$$\widehat{L}_{\dot{x}}(t_j)=(-1)^j\widehat{l}_{x_j},\ j=0,1 \iff p(t_j)=(-1)^j\widehat{l}_{x_j},\ j=0,1. \qquad (3c)$$

PROOF. Problem (P_3) reduces to the smooth problem with equality-type constraints considered above. Indeed, by setting $\mathcal{F}(z)(t) = \dot{x}(t) - \varphi(t, x(t), u(t))$, $\mathcal{G}(z) = (\mathcal{B}_1(z), \ldots, \mathcal{B}_m(z))$, $F(z) = (\mathcal{F}(z), \mathcal{G}(z))$ and $f_0(z) = \mathcal{B}_0(z)$, problem (P_3) can be written as

$$f_0(z) \to \min, \quad F(z) = 0. \tag{P_2'}$$

The smoothness assumptions imply those in Corollary 2 (see Lemmas 2–4). Closedness of the image follows from Lemmas 5 and 6. Hence the stationarity conditions (2) are fulfilled; i.e, there are Lagrange multipliers $\mu_0, \mu_1, \ldots, \mu_m$ and $\nu \in Y^*$, where $Y = C(\Delta, \mathbb{R}^n)$, not equal to zero simultaneously such that

$$\widehat{\mathcal{L}}_{x(\cdot)} = 0 \iff \int_{t_0}^{t_1} \widehat{g}_x(t) \cdot x(t)\, dt + \widehat{l}_{\xi_0} \cdot x(t_0) + \widehat{l}_{\xi_1} \cdot x(t_1) + \langle \nu, y(\cdot) \rangle = 0 \tag{i}$$

for any $x(\cdot) \in C^1(\Delta, \mathbb{R}^n)$ and $x_0 \in \mathbb{R}^n$ such that $\dot{x}(t) - \widehat{\varphi}_x(t)x(t) = y(t)$, $x(t_0) = x_0$, and

$$\widehat{\mathcal{L}}_{u(\cdot)} = 0 \iff \int_{t_0}^{t_1} \widehat{g}_u(t) \cdot u(t)\, dt + \langle \nu, -\widehat{\varphi}_u(\cdot)u(\cdot) \rangle = 0 \tag{ii}$$

for any $u(\cdot) \in C(\Delta, \mathbb{R}^r)$.

Let $p(\cdot)$ be a solution to the equation

$$-\dot{p} = p\widehat{\varphi}_x(t) - \widehat{g}_x(t), \quad p(t_1) = -\widehat{l}_{\xi_1}. \tag{iii}$$

Then taking into account (iii) and the expression for $\dot{x}(\cdot)$ in (i), we have

$$p(t_1) \cdot x(t_1) - p(t_0) \cdot x(t_0) = \int_{t_0}^{t_1} \frac{d}{dt}(p(t) \cdot x(t))$$

$$= \int_{t_0}^{t_1} (\dot{p}(t) \cdot x(t) + p(t) \cdot \dot{x}(t))\, dt$$

$$= \int_{t_0}^{t_1} (-p(t)\widehat{\varphi}_x(t)x(t) + \lambda_0 \widehat{g}_x(t) \cdot x(t) + p(t)\widehat{\varphi}_x(t)x(t) + p(t) \cdot y(t))\, dt$$

$$= \int_{t_0}^{t_1} (\widehat{g}_x(t) \cdot x(t) + p(t) \cdot y(t))\, dt.$$

Together with (i) this implies that

$$-\int_{t_0}^{t_1} p(t) \cdot y(t)\, dt = (\widehat{l}_{\xi_0} - p(t_0)) \cdot x(t_0) + \langle \nu, y(\cdot) \rangle$$

for any $(x_0, y(\cdot)) \in \mathbb{R}^n \times C(\Delta, \mathbb{R}^n)$; hence

$$\langle \nu, y(\cdot)\rangle = \int_{t_0}^{t_1} p(t) \cdot y(t)\, dt, \quad p(t_0) = \widehat{l}_{\xi_0}. \tag{iv}$$

Now (ii) and (iii) imply that

$$p(t)\widehat{\varphi}_u(t) = \widehat{g}_u(t). \tag{v}$$

The equalities $(iii) - (v)$ contain all the assertions of Corollary 3. □

REMARK. 1. For the sake of simplicity we considered here the problem with fixed points. The same arguments slightly modified allow for obtaining necessary conditions for extremum in the general setting.

2. The origination of the calculus of variations is attributed to I. Bernoulli, who posed the problem on brachistochrone in 1696. The first necessary conditions for extremum in a general problem of this type, the so-called simplest problem of the calculus of variations,

$$\int_a^b f(x, y(x), y'(x))\, dx \to \text{extr}, \quad y(a) = y_0,\ y(b) = y_1,$$

were obtained by Euler in 1714. Afterwards various generalizations of the simplest problem were considered: the isoperimetric problem, the Lagrange, Bolza, Mayer problems, etc. Necessary conditions for extremum in these problems were obtained by Euler, Lagrange, Gauss, Ostrogradskii, Poisson, Kneser, Mayer, and many others. Practically all the cases reduce either to Corollary 2 or to Corollary 3.

4. *Smooth problems with equalities and inequalities.*

COROLLARY 4 (the Lagrange principle for smooth problems with equalities and inequalities). *Let X and Y be Banach spaces and U a neighborhood of a point $\widehat{x} \in X$. Further, let functions $f_i \colon U \to \mathbb{R}$, $i = 0, 1, \dots, m$, and a mapping $\mathcal{F} \colon U \to Y$ be strictly differentiable at $\widehat{x}$ and $\operatorname{Im} \mathcal{F}'(\widehat{x}) = Y$. If $\widehat{x}$ affords a local minimum in the problem with equality and inequality constraints*:

$$f_0(x) \to \min, \qquad f_i(x) \le 0, \quad i = 1, \dots, m, \quad \mathcal{F}(x) = 0, \tag{P_4}$$

then there exist Lagrange multipliers $\mu_i \in \mathbb{R}$, $i = 0, 1, \dots, m$, and $\lambda \in Y^*$ not equal to zero simultaneously such that the Lagrange function

$$\mathcal{L}(x, \mu_0, \dots, \mu_m, \lambda) = \sum_{i=0}^{m} \mu_i f_i(x) + \langle \lambda, \mathcal{F}(x)\rangle$$

satisfies the conditions:

— *nonnegativity,*

$$\mu_i \geq 0, \quad i = 0, 1, \ldots, m, \tag{4a}$$

— *complementary slackness,*

$$\mu_i f_i(\widehat{x}) = 0, \quad i = 1, \ldots, m, \tag{4b}$$

— *and stationarity,*

$$\mathcal{L}_x(\widehat{x}, \mu_0, \ldots, \mu_m, \lambda) = 0. \tag{4c}$$

PROOF. If $\operatorname{Im} \mathcal{F}'(\widehat{x}) \neq Y$, then by Lemma 1 there exists a functional $\lambda \in Y^*$ such that $\langle \lambda, \mathcal{F}'(\widehat{x})x \rangle = 0$ for any $x \in X$. Setting $\mu_i = 0$, $i = 0, 1, \ldots, m$, we obtain that conditions $(4a)$, $(4b)$ and $(4c)$ are fulfilled.

Let $\operatorname{Im} \mathcal{F}'(\widehat{x}) = Y$. Rewrite the problem (P_4) as

$$f_0(x) \to \min, \qquad F(x, u) = 0, \tag{P_4'}$$

where $F\colon U \times \mathbb{R}^m_+ \to Y \times \mathbb{R}^m$, $F(x,u) = (\mathcal{F}(x), f_1(x) + u_1, \ldots, f_m(x) + u_m)$. Clearly $\widehat{x}$ is a local minimum point in (P_4) if and only if $(\widehat{x}, 0)$ is a strong local minimum in (P_4'). Applying the smooth-convex principle (the Banach-space, smoothness, and convexity properties are easily verified; finiteness of codimension follows from Lemma 5), we obtain that $(4c)$ follows from the stationarity condition, while $(4a)$ and $(4b)$ follow from the maximum principle. $\square$

REMARK. The finite-dimensional version of this corollary was proved by John (1948), and the infinite-dimensional version by Dubovitskii and Milyutin (1965).

5. *Lyapunov's problems.*

COROLLARY 5 (the Lagrange principle for Lyapunov's problems). *Let Δ be a (finite or infinite) interval of the real line, let U be a measurable subset of $\mathbb{R}^r$, and let $f_i\colon \Delta \times U \to \mathbb{R}$, $0 \leq i \leq m$, be continuous functions. Let Z be a vector space, and suppose that functions $g_i\colon Z \to \mathbb{R}$ are convex for $0 \leq i \leq m'$ and affine for $m' + 1 \leq i \leq m$. Let A be a convex subset of Z and $\mathcal{V}$ the class of all measurable mappings from Δ to U. If an element $(\widehat{u}(\cdot), \widehat{z}) \in \mathcal{V} \times A$ affords the absolute* *minimum in the Lyapunov problem*

$$\int_\Delta f_0(t,u(t))\,dt + g_0(z) \to \min,$$
$$\int_\Delta f_i(t,u(t))\,dt + g_i(z) \le 0, \quad i=1,\dots,m', \tag{P_5}$$
$$\int_\Delta f_i(t,u(t))\,dt + g_i(z) = 0, \quad i=m'+1,\dots,m, \quad z\in A,$$

then there are Lagrange multipliers $\lambda_i \in \mathbb{R}$, $i=0,1,\dots,m$, not equal to zero simultaneously such that the Lagrange function

$$\mathcal{L}((u(\cdot),z),\lambda_0,\dots,\lambda_m) = \int_\Delta L(t,u(t))\,dt + l(z),$$

where $L=\sum_{i=0}^m \lambda_i f_i$ and $l=\sum_{i=0}^m \lambda_i g_i$, satisfies the following conditions:

— *nonnegativity,*

$$\lambda_i \ge 0, \quad i=0,1,\dots,m, \tag{5a}$$

— *complementary slackness,*

$$\lambda_i\left(\int_\Delta f_i(t,\widehat{u}(t))\,dt + g_i(\widehat{z})\right) = 0, \quad i=1,\dots,m', \tag{5b}$$

— *and the maximum principle,*

$$\min_{u\in U} L(t,u) = L(t,\widehat{u}(t)) \text{ for almost all } t\in\Delta, \quad \min_{z\in A} l(z) = l(\widehat{z}). \tag{5c}$$

PROOF. We reduce problem (P_5) to the smooth-convex problem (P) by setting $X=\mathbb{R}$, $\mathcal{U}=\mathcal{V}\times A\times\mathbb{R}_+^{m+1}$, $u=(u(\cdot),z,\alpha)$, $\alpha=(\alpha_0,\dots,\alpha_m)$, $f_0(x)=x$, $F(x,u)=(\varphi_0(x,u),\varphi_1(u),\dots,\varphi_m(u))$, where $\varphi_0(x,u)=\int_\Delta f_0(t,u(t))\,dt+g_0(z)-x+\alpha_0$ and $\varphi_i(u)=\int_\Delta f_i(t,u(t))\,dt+g_i(z)+\alpha_i$, $1\le i\le m$. It is elementary to verify (taking into account Lyapunov's theorem) that the conditions of the Lagrange principle for smooth-convex problems are fulfilled. An application of this theorem immediately leads to the desired result.□

Bibliography

N. I. Akhiezer (1965), *Lectures on the theory of approximation*, Nauka, Moscow; English transl., Frederik Ungar, New York, 1956.

V. M. Alekseev, E. M. Galeev, and V. M. Tikhomirov (1984), *Optimization problem book*, Nauka, Moscow (Russian); French transl., *Recueil de problèmes d'optimisation*, Mir, Moscow, 1987.

V. M. Alekseev, V. M. Tikhomirov, and S. V. Fomin (1979), *Optimal control*, Nauka, Moscow; English transl., Consultants Bureau, New York, 1987.

V. V. Arestov (1989), *Best recovery of operators and related problems*, Trudy Mat. Inst. Steklov **189**, 3–20; English transl. in Proc. Steklov Inst. Math. **189** (1991).

______ (1996), *Approximation of unbounded operators by bounded operators and related extremal problems*, Uspekhi Mat. Nauk **51**, no. 6, 89–124; English transl. in Russian Math. Surveys **51**.

T. Bonnesen and W. Fenchel (1934), *Theorie der konvexen Körper*, Springer-Verlag, Berlin.

Yu. D. Burago and V. A. Zalgaller (1988), *Geometric inequalities*, Springer-Verlag, Berlin.

G. B. Dantzig (1963), *Linear programming and extensions*, Princeton Univ. Press, Princeton.

A. Ya. Dubovitskii and A. A. Milyutin (1965), *Problems for extremum under constraints*, Zh. Vychisl. Mat. i Mat. Fiz. **5**, no. 3, 395–453; English transl. in USSR Comput. Math. and Math. Phys. **5** (1965).

N. Dunford and J. T. Schwartz (1958), *Linear operators.* Vol. I, Interscience, New York.

I. Ekeland and R. Teman (1976), *Convex analysis and variational problems*, SIAM, New York.

J. Farkas (1901), *Über die Theorie der einfachen Ungleichungen*, J. Reine und Angew. Math. **124**, 1–24.

W. Fenchel (1951), *Convex cones, sets and functions*, Princeton Univ. Press, Princeton.

A. T. Fuller (1960), *Relay control systems optimized by various performance criteria*, Proc. First World Congress IFAC, Moscow, vol. 1, London, pp. 510–519.

V. N. Gabushin (1969), *On the best approximation of the differentiable operator on the half-line*, Mat. Zametki **6**, no. 5, 573–582; English transl. in Math. Notes **6**.

D. Gale (1960), *The theory of linear economic models*, McGraw-Hill, New York.

J. B. Giles (1982), *Convex analysis with applications in differentiation of convex functions*, Pitman and Hall, London.

J. Hadamard (1914), *Sur le module maximum d'une fonction et de ses dérivées*, Bull. Soc. Math. France, C. R. de Séances **42**, 68–72.

G. G. Hardy, J. E. Littlewood, and G. Pólya (1934), *Inequalities*, Cambridge Univ. Press, Cambridge.

E. Helly (1923), *Über Mengen konvexer Körper mit gemeinschatlichen Punkten*, Über. Deutsch. Math. Verein. **32**, 175–176.

R. B. Holmes (1975), *Geometric functional analysis and its applications*, Springer-Verlag, New York–Heidelberg–Berlin.

A. D. Ioffe and V. M. Tihomirov [Tikhomirov] (1974), *Theory of extremal problems*, Nauka, Moscow; English transl., North-Holland, Amsterdam, 1979.

F. John (1948), *Extreme problems with inequalities as subsidary conditions*, Studies and essays presented to R. Courant on his 60th birthday, January 8, 1948, Intersociety, New York, pp. 187–204.

L. V. Kantorovich (1939), *Mathematical methods of production management and planning*, Leningrad Univ., Leningrad. (Russian)

——— (1942), *On transportation of masses*, Dokl. Akad. Nauk SSSR **37**, 227–229. (Russian)

L. V. Kantorovich and G. Sh. Rubinstein (1957), *On a functional space and some extremal problems*, Dokl. Akad. Nauk SSSR **115**, no. 6, 1058–1061. (Russian)

A. Khovansky, D. Bernstein, and A. Kushnirenko (1976), *Mixed volumes of Newton polynomials, Milnor's numbers and numbers of solutions of algebraic equations*, Uspekhi Mat. Nauk **32**, no. 3, 196. (Russian)

A. N. Kolmogorov (1939), *Inequalities between the suprema of consecutive derivatives of an arbitrary function on an infinite interval*, Uchen. Zap. Moskov. Univ. Mat. **3**, no. 30, 3–16; English transl. in: Selected works of A. N. Kolmogorov, Vol. 1: Mathematics and Mechanics, Kluwer, Dordrecht, 1991, pp. 277–290.

A. Koushnirenko (1976), *Polyhedres de Newton et nombre de Milnor*, Inv. Math. **32**, 1–31.

N. P. Kuptsov (1975), *Kolmogorov estimates for derivatives in* $L_2(0,\infty)$, Trudy Mat. Inst. Steklov **138**, 94–117; English transl. in Proc. Steklov Inst. Math. **138**.

S. S. Kutateladze (1979), *Convex operators*, Uspekhi Mat. Nauk **34**, no. 1, 167–196. (Russian)

Ky Fan (1956), *On systems of linear inequalities*, Linear Inequalities and Related Systems, Princeton Univ. Press, Princeton, NJ.

E. Landau (1913), *Ungleichungen für zweimal differenzierbar Funktionen*, Proc. London Math. Soc. **2**, no. 13, 43–49.

V. L. Levin (1978), *The Monge–Kantorovich problem on transportation of masses*, Matematicheskaya Ekonomika i Funkcional'nyi Analis (Mathematical Economics and Functional Analysis), Nauka, Moscow, pp. 94–108. (Russian)

——— (1985), *Convex analysis in spaces of measurable functions and its applications in mathematics and economics*, Nauka, Moscow. (Russian)

V. L. Levin and A. A. Milyutin (1979), *The problem on transportation of masses with discontinuous cost function and the duality of convex extremal problems in terms of masses*, Uspekhi Mat. Nauk **34**, no. 3, 3–68. (Russian)

Yu. I. Lyubich (1960), *Inequalities between powers of a differential operator*, Izv. Akad. Nauk SSSR Ser. Mat. **24**, 825–864. (Russian)

L. A. Lyusternik (1934), *On conditional extrema of functionals*, Mat. Sbornik **41**, no. 3, 390–401. (Russian)

——— (1956), *Convex figures and polyhedra*, Gostekhizdat, Moscow. (Russian)

G. G. Magaril-Il'yaev (1983), *Inequalities for derivatives and duality*, Trudy Math. Inst. Steklov **161**, 183–194. (Russian)

G. G. Magaril-Il'yaev and K. Yu. Osipenko (1991), *Optimal recovery of functionals based on inaccurate data*, Mat. Zametki **50**, no. 1, 85–93; English transl. in Math. Notes **50**, 1274–1279.

——— (2002), *Optimal recovery of functions and their derivatives from Fourier coefficients prescribed with an error*, Mat. Sbornik **193**, no. 3, 79–100; English transl. in Sbornik: Mathematics **193**, no. 3, 387–407.

——— (2003), *Optimal recovery of functions and their derivatives from approximate information on the spectrum and inequalities for derivatives*, Funktsional Anal. i Prilozhen. **37**, no. 3. (Russian)

G. G. Magaril-Il'yaev and V. M. Tikhomirov (1981), *Kolmogorov's inequality for fractional derivatives on a half-line*, Analysis Math. **7**, no. 1, 37–47. (Russian)

——— (1997), *Kolmogorov-type inequalities for derivatives*, Mat. Sbornik **188**, no. 12, 73–106; English transl. in Sbornik: Mathematics **188**, no. 12, 1799–1832.

——— (2000), *Convex analysis and its applications* (2nd ed. 2003), Editorial URSS, Moscow. (Russian)

C. A. Micchelli and T. J. Rivlin (1977), *A survey of optimal recovery*, Optimal Estimation in Approximation Theory, Plenum, New York, pp. 1–54.

——— (1985), *Lectures on optimal recovery*, Lecture Notes in Math., vol. 1129, Springer-Verlag, Berlin, pp. 21–93.

H. Minkowski (1896), *Geometrie der Zahlen*, Teubner, Leipzig.

G. Monge (1784), *Mémoire sur la théorie des déblais et des remblais*, Histoire de l'Académie Royale des Science, 666–704.

J. J. Moreau (1962), *Fonctions convexes en dualité*, Fac. des Sciences de Montpellier. Sem. de Mathématiques.

K. Yu. Osipenko (1976), *Best approximation of analytic functions from their values at a finite number of points*, Mat. Zametki **19**, no. 1, 29–40; English transl. in Math. Notes **19**, 17–23.

——— (2000), *Optimal recovery of analytic functions*, Nova Science Publ., Huntington, NY.

R. R. Phelps (1978), *Differentiability of convex functions on Banach spaces*, Saunders, London.

L. Plaskota (1996), *Noisy information and computational complexity*, Cambridge Univ. Press, Cambridge.

J. Radon (1921), *Körper, die einen gemeinsamen Punkt enthalten*, Math. Ann. **83**, 113–115.

R. T. Rockafellar (1970), *Complex analysis*, Princeton Univ. Press, Princeton, NJ.

A. Sard (1949), *Best approximate integration formulae; best approximation formulae*, Amer. J. Math. **71**, 80–91.

S. A. Smolyak (1965), *Optimal recovery of functions and functionals thereof*, Kandidat Thesis, Moscow. (Russian)

B. Sz.-Nagy (1941), *Über Integralungleichungen zwischen einer Function und ihrer Ableitung*, Acta Sci. Math. **10**, 64–74.

L. N. Taikov (1968), *Kolmogorov-type inequalities and the best formulae for numerical differentiation*, Mat. Zametki **4**, no. 2, 233–238; English transl. in Math. Notes **4**.

V. M. Tikhomirov and G. G. Magaril-Il'yaev (1991), *Inequalities for derivatives*, Selected works of A. N. Kolmogorov, Vol. 1: Mathematics and Mechanics, Kluwer, Dordrecht, pp. 442–446.

H. Weyl (1935), *Elementare Theorie der konvexen Polyheder*, Comm. Math. Helv. **7**, 290–306.

M. I. Zelikin and V. F. Borisov (1994), *Theory of chattering control*, Birkhäuser, Boston.

Index